Inoculating Cities

Inoculating Cities
Case Studies of Urban Pandemic
Preparedness

Edited by

Rebecca Katz

Matthew Boyce

ELSEVIER

ACADEMIC PRESS
An imprint of Elsevier

Academic Press is an imprint of Elsevier
125 London Wall, London EC2Y 5AS, United Kingdom
525 B Street, Suite 1650, San Diego, CA 92101, United States
50 Hampshire Street, 5th Floor, Cambridge, MA 02139, United States
The Boulevard, Langford Lane, Kidlington, Oxford OX5 1GB, United Kingdom

Notices
Knowledge and best practice in this field are constantly changing. As new research and experience broaden our understanding, changes in research methods, professional practices, or medical treatment may become necessary.

Practitioners and researchers must always rely on their own experience and knowledge in evaluating and using any information, methods, compounds, or experiments described herein. In using such information or methods they should be mindful of their own safety and the safety of others, including parties for whom they have a professional responsibility.

To the fullest extent of the law, neither the Publisher nor the authors, contributors, or editors, assume any liability for any injury and/or damage to persons or property as a matter of products liability, negligence or otherwise, or from any use or operation of any methods, products, instructions, or ideas contained in the material herein.

Library of Congress Cataloging-in-Publication Data
A catalog record for this book is available from the Library of Congress

British Library Cataloguing-in-Publication Data
A catalogue record for this book is available from the British Library

ISBN: 978-0-12-820204-3

For information on all Academic Press publications visit our website at
https://www.elsevier.com/books-and-journals

Publisher: Andre Gerhard Wolff
Acquisitions Editor: Kattie Washington
Editorial Project Manager: Sam W. Young
Production Project Manager: Selvaraj Raviraj
Cover Designer: Mark Rogers

Typeset by TNQ Technologies

Working together
to grow libraries in
developing countries

www.elsevier.com • www.bookaid.org

Contents

Contributors

Emmanuel Agogo, Directorate of Prevention Programmes and Knowledge Management, Nigeria Centre for Disease Control, Abuja, Nigeria

Aditya Ajith, Chief Minister's Urban Leaders Fellow, Government of NCT of Delhi, New Delhi, India

Robin Albrandt, Clark County Public Health, Vancouver, WA, United States

Sara M. Allinder, Center for Innovation in Global Health, Georgetown University, Washington, DC, United States

Adejare (Jay) Atanda, School of Community Health & Policy, Morgan State University, Baltimore, MD, United States

Matthew Boyce, Center for Global Health Science & Security, Georgetown University, Washington, DC, United States

Elliot Brennan, Myanmar Health & Development Consortium, Yangon, Myanmar

Hank J. Brightman, College of Maritime Operational Warfare/Civilian-Military Humanitarian Response Program, U.S. Naval War College, Newport, RI, United States

Nicholas Cagliuso, Emergency Management, NYC Health + Hospitals, New York, NY, United States

Anna M. Carter, Center for Innovation in Global Health, Georgetown University, Washington, DC, United States

Chioma Dan-Nwafor, Directorate of Surveillance and Epidemiology, Nigeria Centre for Disease Control, Abuja, Nigeria

Priya Dhagat, Emergency Management, System-wide Special Pathogens Program, NYC Health + Hospitals, New York, NY, United States

Myles Druckman, Global Health Services, International SOS, Los Angeles, CA, United States

Kayode Fasominu, Volte health Systems, Abuja, Nigeria

Brian Gerber, Watts College of Public Service & Community Solutions, Arizona State University, Phoenix, United States

Samayita Ghosh, Centre for Environmental Health, Public Health Foundation of India, Gurgaon, Haryana, India

Philippe Guibert, Europe Health Consulting, International SOS, Paris, France

David James Heslop, School of Public Health, University of New South Wales, Sydney, NSW, Australia

Charles B. Holmes, Center for Innovation in Global Health, Georgetown University, Washington, DC, United States

Chikwe Ihekweazu, Office of the Director General, Nigeria Centre for Disease Control, Abuja, Nigeria

Elsie Ilori, Directorate of Surveillance and Epidemiology, Nigeria Centre for Disease Control, Abuja, Nigeria

Rebecca Katz, Center for Global Health Science & Security, Georgetown University, Washington, DC, United States

Gift Kawalazira, Blantyre District Health Office, Government of Malawi, Blantyre, Malawi

Irene Lai, Medical Information and Analysis, International SOS, Sydney, NSW, Australia

Folake Lawal, Medical College of Georgia, Augusta University, Augusta, GA, United States

Sandii Lwin, Myanmar Health & Development Consortium, Yangon, Myanmar

Chimwemwe Mablekisi, The Malawi National AIDS Commission, Lilongwe, Malawi

Raina Chandini MacIntyre, The Kirby Institute, University of New South Wales, Sydney, NSW, Australia

Syra Madad, Emergency Management, System-wide Special Pathogens Program, NYC Health + Hospitals, New York, NY, United States

Shyamala Mani, Centre for Environmental Health, Public Health Foundation of India, Gurgaon, Haryana, India

Alan Melnick, Clark County Public Health, Vancouver, WA, United States

Kyi Minn, Myanmar Health & Development Consortium, Yangon, Myanmar

Takako Misaki, Division of Planning and Management, Kawasaki City Institute for Public Health, Kawasaki, Kanagawa, Japan

Samuel Mutbam, Nigeria Country Office, World Health Organization, Abuja, Nigeria

William Nwachukwu, Directorate of Surveillance and Epidemiology, Nigeria Centre for Disease Control, Abuja, Nigeria

Adesola Ogunsola, Directorate of Surveillance and Epidemiology, Nigeria Centre for Disease Control, Abuja, Nigeria

Nobuhiko Okabe, Director General, Kawasaki City Institute for Public Health, Kawasaki, Kanagawa, Japan; Division of Planning and Management, Kawasaki City Institute for Public Health, Kawasaki, Kanagawa, Japan

Ifeanyi Okudo, Nigeria Country Office, World Health Organization, Abuja, Nigeria

Oyeladun Okunromade, Directorate of Surveillance and Epidemiology, Nigeria Centre for Disease Control, Abuja, Nigeria

Oyeronke Oyebanji, Office of the Director General, Nigeria Centre for Disease Control, Abuja, Nigeria

Rachael Piltch-Loeb, Emergency Preparedness Research Evaluation & Practice Program, Harvard TH Chan School of Public Health, Boston, MA, United States

Saskia Popescu, Schar School of Policy and Government - Biodefense, George Mason University, Arlington, VA, United States

Poornima Prabhakaran, Centre for Environmental Health, Public Health Foundation of India, Gurgaon, Haryana, India

Tomoya Saito, Center for Emergency Preparedness and Response, National Institute of Infectious Diseases, Shinjuku-ku, Tokyo, Japan

Ibrahim Seriki, Zamafara State Field Office, World Health Organization, Zamfara, Nigeria

Richa Sharma, Centre for Environmental Health, Public Health Foundation of India, Gurgaon, Haryana, India

Amy Simpson, Medical Information and Analysis, International SOS, Sydney, NSW, Australia

Tyler R. Smith, Cooper/Smith, Washington, DC, United States

Michael A. Stoto, Department of Health Systems Administration, Georgetown University, Washington, DC, United States

Francesca Viliani, Public Health, International SOS, Copenhagen, Denmark

Kyaw San Wai, Myanmar Health & Development Consortium, Yangon, Myanmar

Roxanne Wolfe, Clark County Public Health, Vancouver, WA, United States

Acknowledgments

The editors of this volume were supported by a grant from the Open Philanthropy Project to the Georgetown University Center for Global Health Science and Security.

Introduction: cities, infectious disease, and the local governance of health security

Cities are cultural, economic, academic, and historic hubs and a relatively recent part of human history. For much of their evolutionary history, human populations lived in small, nomadic groups and populations remained relatively stable as a result of comparable reproduction and mortality rates [1]. But some 12 thousand years ago, during the Agricultural Revolution, humans abandoned this more nomadic lifestyle, formed more permanent settlements, and began to farm the land. This dramatic anthropologic shift led to rapid increases in human population size and population density and eventually the first city—Çatalhöyük, a densely populated human settlement of roughly 8000 individuals that emerged around 7000 B.C. in modern-day Turkey [2]. Not long after, much larger urban settlements were established, with some growing as large as 50,000 individuals by 3000 B.C. [3].[1]

This demographic trend—urbanization—has not let up since and cities have emerged all over the world—everywhere from Cairo to Calgary to Canberra. However, over the past 250 years, our world has urbanized at an unprecedented rate. Largely catalyzed by the Industrial Revolution that began in the mid-18th century, people flocked to cities to pursue the promises of economic prosperity and other advantages not available in rural areas. Many cities in Western Nations experienced rapid increases in population, such as Chicago, which grew from just under 30,000 in 1850 to nearly 2.2 million in 1910 [4]. Presently, many cities in Africa and Asia are experiencing analogous population booms. Dar es Salaam, Tanzania, has grown from 67,000 in the 1950 to approximately 3.4 million in 2010 and is projected to reach over 5.6 million by 2025 [5]. Similarly, Guangzhou, China, has grown from approximately 2.5 million people in 1950 to 12.7 million in 2010 [6,7].

These two more recent examples do well to illustrate broader trends in urbanization. According to the United Nations, in 1990, 43 percent (2.3 billion) of the world's population lived in urban areas; by 2015, this had grown to 54

---

1. The City of Uruk located in present day Iraq is considered by some to be the 'first city.' But because there is no universally accepted definition for what constitutes a city (as discussed later), this claim is contested.

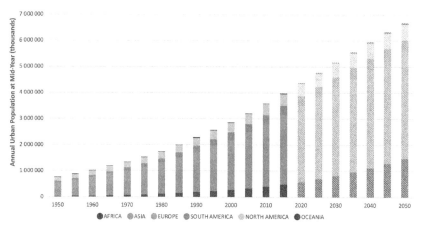

FIGURE 1   Global urban population at mid-year, 1950–2050 [10]. Visualized data are sourced from the United Nations population division [9].

percent (4 billion) [8]; and estimates suggest that by 2050, cities will host an additional 2.5 billion people and nearly 70 percent of the world's population— with a majority of this growth occurring in African and Asian cities [9] (Fig. 1).

Defining urban

Urbanization refers to an increase in the movement and settling of people in urban areas [11]. However, the word "urban" does not have a universal accepted definition, and various interpretations are founded on a range of understandings and factors. Some definitions are based on physical location, such as that used by the United Nations, which delineates urban areas as relating to the city proper, the urban agglomeration, and the metropolitan area [9]. The city proper is meant to describe the administrative boundary of a city; the urban agglomeration accounts for the larger, adjacent areas; and the metropolitan area is used to describe the greater area that has strong economic or social ties to the city proper.

Other classifications define urban areas by economic activity or output. For example, the Japanese Government states that for an area to be considered urban, at least 60 percent of the population must be engaged in manufacturing, trade, or other urban type of business [9]. In contrast to specifying that the economy must be predominantly manufacturing or trade-based, other countries, such as Botswana, Croatia, and India, specify that in order to be urban, a majority of the local economy must be nonagricultural [9].

Some countries' definitions use total population or population density as the hallmark characteristic of urban areas. However, as alluded to previously, the specific requirements vary considerably. For example, the urban threshold

TABLE 1 Variations in the minimum population size required to be considered urban [9].

Threshold	Country
200	*Denmark, Iceland, Sweden*
400	*Albania*
600	*Lao People's Democratic Republic*
1000	*Dominica, Fiji, Micronesia, New Zealand, San Marino*
1500	*Somalia*
2000	*Argentina, Bolivia, Colombia, Czechia, Eritrea, Ethiopia, Guadeloupe, Kenya, Liberia, Luxembourg, Martinique, Norway, Portugal, Sierra Leone, Uganda*
2500	*Ireland, Mexico, Venezuela*
3000	*Central African Republic, Gabon*
4000	*Vietnam*
5000	*Belgium, Comoros, Gambia, Ghana, Jordan, Lebanon, Madagascar, Mauritania, Qatar, Saudi Arabia, Slovakia, South Sudan, Sudan, Tunisia*
10,000	*Australia, Benin, Italy, Kuwait, Senegal, Solomon Islands, Spain, Switzerland, United Kingdom*
20,000	*The Netherlands, Nigeria, Syrian Arab Republic*
30,000	*Mali*

in Denmark requires that localities have at least 200 inhabitants, while the threshold in Ethiopia requires 2000 inhabitants or more, and the threshold in Australia requires a minimum of 10,000 inhabitants (Table 1). The United States' definition of urban includes both a total population threshold and a population density requirement.

Other interpretations of urbanicity use a combination of these factors, such as that used by Côte d'Ivoire, which defines urban as areas with at least 10,000 inhabitants or areas with between 4000 and 10,000 inhabitants and with more than 50 percent of households engaged in nonagricultural activities [9]. And for some areas, the requirements are more opaque. For example, Malawi defines urban as town planning areas and district centers, whereas Myanmar does not have an official definition [9]. For the purposes of this book, we adhere to self-definitions of what constitutes a city.

Cities and infectious diseases

Regardless of the definition, the larger trend of urbanization has resulted in a situation in which urban health sits squarely at the forefront of public health. But this reality is not without precedent, as cities have a storied history with infectious diseases. In 430 B.C., the ancient city of Athens experienced an epidemic—hypothesized to be everything from tuberculosis to Ebola to smallpox—that killed approximately one-third of the Athenian population [12]. Centuries later, when the Black Death swept across Europe, Asia, and North Africa, the effects were acutely felt by cities, some of which lost 50 percent of their populations to the pandemic. More recently, severe acute respiratory syndrome (SARS) caused alarm when it emerged in southern China in 2002 and then quickly spread around the world—primarily in dense urban areas including Hong Kong, Singapore, Toronto, and Hanoi [13].

An important question to ask is why are the horrors of outbreaks so well chronicled in urban areas? Is it that cities act as the cultural centers of our world and thus receive more attention from the media, politicians, and scholars; or is it that there are particular characteristics that make urban areas conducive to infectious disease outbreaks? We posit it is both.

Cities contain characteristics that can promote the spread of infectious diseases both within and between cities. Frameworks for classifying these risk factors further categorize them as those relating characteristics of urban populations, those relating to the physical environments of cities, and those relating to social determinants of health in urban areas [14].

High population densities are common to cities due to the combination of natural population growth—albeit, generally with lower birth rates than rural regions—and in-migration and represent one example of a unique urban risk relating to urban populations. As economist Edward Glaeser has written, "the same [population] density that spreads ideas can spread disease" [15]. This is because the dense urban populations can provide conditions that promote disease emergence and transmission [16]—especially diseases with respiratory and oral—fecal transmission pathways [14]—which can compound the prevention and control of infectious diseases. Indeed, it is because of urbanization that human populations are large enough to maintain diseases such as measles in endemic form [17]. This is also reflected in individual-level characteristics of populations. Research has demonstrated that populations with a longer history of urban residence are better genetically adapted to resisting respiratory infections—supporting the assertion that these diseases became an increasingly important cause of human mortality after the advent of urbanization and highlighting the importance of population density in determining human health [18].

Furthermore, what was an endemic disease in one population could be the source of an epidemic in another population. This is especially true for migrants who often flock to cities in search for better lives and economic opportunity.

This population presents two important considerations for urban health: the introduction of new diseases and increased susceptibility to endemic diseases. Other examples of risk factors relating to urban population characteristics include vaccination rates, personal behaviors (e.g., handwashing, condom use, etc.), and cultural norms.

Risk factors relating to the physical environment of cities concern both microlevel characteristics, such as access to clean water or transport networks, and more macrolevel considerations such as altitude and climate. In today's highly globalized world, cities act as transportation hubs in highly connected networks, and the presence of large airports, seaports, and train and bus stations in cities facilitates mass movements of people and goods [19,20]. This is important for the epidemiology of infectious diseases, as most epidemics and pandemics have spread following transportation, commercial, and traveling networks.

If cities are conceptualized as nodes in a network, network theory can be used to frame this discussion. Theory suggests that two characteristics have the potential to greatly affect the spread of infectious disease: transitivity (the propensity for clustering within a network) and centrality (the significance of a node is within a network) [21,22]. Highly transitive nodes—that is, those that can be reached multiple times via different mechanisms—promote disease because they act as bridges for disease transmission by inherently providing more opportunities for infection. For example, cities on the eastern seaboard of the United States, such as New York, Philadelphia, Baltimore, and Washington, D.C., represent a transitive network with each of these cities linked to one another via multiple routes.

The centrality of nodes, in this case, can also be thought of as a node with many contacts to other notes or a high degree of distribution. This would account for why an infectious disease outbreak emerging in Dubai is at greater risk for global spread compared to an outbreak emerging in Dayton. Nodes with a high degree of centrality are prone to not only catalyzing facilitating the global spread of disease but must also have strong public health systems in place to address the higher risk for imported disease. Taken together, these considerations ultimately render cities as both a rate-enhancing and rate-limiting factor in the global spread of infectious disease.

The presence of animal wholesale and retail markets represents another common physical characteristic in cities and one that can greatly influence the emergence of infectious diseases. These markets are important routes for zoonotic spillover events, or the cross-species transmission of infectious diseases, which can begin a cascade of events that culminate in an epidemic. For example, live poultry markets have been shown to be important in the epidemiology and emergence of avian influenzas such as H5N1 and H7N9 [23−25], and possibly for COVID-19 as well. By some estimates, these kinds of events annually result in billions of cases of human disease and millions of deaths and have resulted in hundreds of billions of dollars in economic losses over the past two decades alone [26].

Social determinants of health and urban inequalities are also important urban determinants of health. Indeed, several researchers have pointed out that living in a city is in itself a social determinant of health [27,28]. Still, within a city, factors such as socioeconomic status, place of residence, race, ethnicity, gender, and education can determine vulnerability to infectious disease outcomes and perpetuate disease transmission. For example, a 14-year study conducted in Taiwan found that lower socioeconomic status was associated with increased risk of more severe disease in patients with dengue fever [29].

Of note is that these risk factors can vary spatially within cities and rarely act in isolation. During the 2002–2003 SARS epidemic that impacted Beijing, Hong Kong, Singapore, and Toronto, population density (a population characteristic) and presence of ports that enabled travel (a physical characteristic) were thought to greatly impact the epidemiology of the outbreak [13,30].

Slums represent another example of a combination of multiple kinds of risk factors. Historically, slums emerged in industrial cities and quickly became focal points for poverty and the spread of disease because of a combination of overcrowding as a result of uncontrolled population growth (a population characteristic) and insufficient access to basic services such as safe housing, drinking water, and adequate sewage facilities (a physical characteristic) [31,32]. Throughout history, slums were sights of well-documented epidemics of smallpox, tuberculosis, typhoid, diphtheria, measles, and yellow fever [33] and have also contributed to the transmission of a variety of other infectious diseases including dengue, chikungunya, Zika, hepatitis, leptospirosis, and cholera [34,35].

The local governance of infectious disease

These unique urban factors make the detection and control of infectious disease outbreaks a direct function of cities, which require robust public health systems. Urban leaders have been grappling with the responsibility of protecting the health of urbanites for centuries. Venice has perhaps the most storied history. In 1377, the city of Dubrovnik issued a decree whereby before entering the city, travelers had to spend 30 days on nearby islands to observe if they would develop disease symptoms. In the midst of a 15th-Century plague outbreak, Venice adapted this by extending the period to 40 days, giving rise to modern concept of the quarantine. The Venetian government also built a public hospital on an island named Santa Maria di Nazareth to reduce the spread of plague by isolating the unwell from healthy populations. The island was commonly called Lazaretum because of its proximity to nearby island dedicated to St. Lazarus, and with time, most began referring to the establishment as Lazzaretto Vecchio, giving rise to the modern concept of a lazaret—a quarantine station for maritime travelers [36].

These types of nonpharmaceutical interventions—including quarantine, isolation, and contract tracing—are often the first actions taken to respond to infectious disease outbreaks and are almost always carried out by municipal public health authorities. But municipal authorities also often have a major role in other response activities including the distribution of medical supplies, supervising vaccination campaigns, managing public communication campaigns, and requesting additional assistance.

The local governance of infectious disease is largely tasked to mayors and local health departments who share the common goal of promoting and protecting well-being within their communities. In many settings, mayors serve as the highest-ranking official in a municipal government. The powers and responsibilities of a mayor can vary widely depending on the larger political system, but as outlined by political scientist Benjamin Barber, mayors have the ability to shape social policy agendas, set budgets, and influence the distribution resources and commodities within cities [37]. Health commissioners are the health department directors for cities and are primarily responsible for overseeing public health activities and initiatives. Together, these two individuals often represent the most trusted sources of public health information during public health emergencies in cities, including infectious disease outbreaks [38].

The size, structure, and specific authorities of these actors and municipal health departments can differ and often depend on the larger political and historical contexts in which they operate [39]. These contexts can also vary widely across a country. For example, in the United States, 29 states have a decentralized organizational model for public health in which local public health agencies are organizationally independent of the state agencies and are primarily governed by local authorities; 6 states have a centralized structure in which public health agencies are directly governed and operated by state governments; 13 states operate using hybrid model; and 2 states do not have any local public health agencies and provide all public health services through state agencies [40]. This results in a situation in which the local governance of disease in Seattle is different from that in Houston, which is different from that in Miami.

Cities with a history of outbreaks or a perceived threat of bioterrorism may be more likely to prioritize public health preparedness. Kingdon's Policy Steams Model is one useful tool for understanding this cycle [41]. Under this model, three independent streams—problems, policy, and politics—converge to focus attention on a specific issue. The problems stream is comprised of all of the issues demanding attention from policymakers; the policy stream is comprised of all of the methods of addressing the problem; and the politics stream represents the tipping point of events and forces that combine to result in action. Central to all of this is a niche filled with policy entrepreneurs and champions who advocate for particular ideas. Mayors, health commissioners, and other administrators in municipal public health departments can and often fill this role for urban public health initiatives.

For example, faced with a rapidly growing measles outbreak in 2019, New York City Mayor Bill de Blasio declared a public health emergency and the health commissioner began a program that mandated vaccination in parts of the city—imposing fines of up to USD 1000 on New Yorkers in the affected neighborhoods who could not prove immunity to the disease or produce a medical exemption and refused to let themselves or their children be vaccinated [42]. This event highlights how when the three streams intersect, with the support of policy champions, meaningful action and governance can occur at the local level within cities.

SARS-CoV-2 and the urban response to the COVID-19 pandemic

Perhaps no recent event demonstrates these concepts better than the rapid, global spread of SARS-CoV-2 and the resulting COVID-19 pandemic. In late December 2019, local hospitals in the megacity (i.e., a city with a population of more than 10 million inhabitants) of Wuhan, China, reported four cases of a "pneumonia of unknown etiology" to the World Health Organization, all linked to a wholesale seafood market. The cases were detected using a surveillance mechanism geared toward the rapid identification of novel pathogens that was established following the 2003 SARS outbreak [43]. Testing quickly revealed that the disease was a novel coronavirus.

The virus then spread throughout China before spreading around the world to Thailand, Japan, Korea, the United States, Singapore, and other countries. Roughly 1 month later, by the end of January, the World Health Organization reported COVID-19 cases in 18 countries on four continents [44]; by the end of February, there were cases in 54 countries on every continent barring Antarctica [45]; and by mid-March, the outbreak was characterized as a pandemic by the Director-General of the World Health Organization [46]. However, this account of the COVID-19 pandemic omits an important detail that from its origins in Wuhan, it was cities that had to grapple with the immediate effects of COVID-19 and launch responses to the public health emergency.

At the time this book was conceptualized, in early 2019, many of the ideas, models, and capacities laid in the more theoretical level—focusing on how they could be applied to future pandemics. However, the past several months have witnessed a transition from theory to reality and the application of many of these ideas and concepts in the response to the COVID-19 pandemic. Indeed, no other event could have validated the importance of urban pandemic preparedness with such emphasis or weight as an actual pandemic. Throughout the pandemic, local leaders have been required to make difficult decisions to restrict travel and economic activity to protect health, use and repurpose existing health infrastructure, and rapidly develop new capacities to mount an effective response to the pandemic.

Some cities have responded to the pandemic well while others have struggled. Such heterogeneity should be expected because of the fantastically diverse set of specific contexts that define the reality of cities and their citizens. Indeed, we have witnessed, in real time, the variability in preparedness, the challenges faced by cities in gathering and sharing both information and best practices, and the necessity of engaging local-level authorities in efforts to prepare and respond to infectious disease outbreaks.

The structure of this book

While the COVID-19 pandemic has certainly underscored the necessity of urban pandemic preparedness and undeniably influenced the collective thinking surrounding preparedness in cities, it is important to acknowledge that not all of the best practices for urban pandemic preparedness are explicitly related to this novel disease. Accordingly, it is crucial to not distort the concept of urban pandemic preparedness and mistake it as synonymous with preparedness for COVID-19 or other respiratory diseases.

This volume uses a wider lens to examine the broader space of urban pandemic preparedness and presents some of the innovative models that are being used to combat infectious disease outbreaks in cities around the world. The chapters consider a diverse set of diseases, geographies, and models in various phases of implementation and have been written by a varied group of authors—including academics, government officials, medical professionals, and others who have had first-hand experiences preparing for and responding to outbreaks in cities—to present a colorful picture of the state of preparedness in cities.

The book is divided into three sections. Part One presents a series of case studies of city-led responses to a variety of recent infectious disease outbreaks, including dengue, Zika, measles, and others. The chapters in Part Two are devoted to examining some of the novel approaches, models, and technologies that are being using to address the risks posed by infectious diseases and prepare cities for outbreaks. Finally, in Part Three, the authors discuss other important considerations for urban public health preparedness. The topics covered in this section include the role of the private sector in promoting urban health security and how cities can prepare for bioterrorism threats.

Chapter 1 presents a case study discussing the experience of controlling an urban outbreak of Dengue in Delhi, India. The authors discuss how the municipality and other urban local bodies are taking measures for dengue control and prevention. These measures include opening fever clinics in the government hospitals, dispensaries, and community clinics, seasonal increases in bed capacity, and an intensive anti-dengue campaign launched to increase awareness and improve control of the disease. They also discuss efforts taken by the Delhi government to form a dengue control unit and the impacts these actions have had on the prevalence of disease.

Chapter 2 focuses on the experiences of a large municipal healthcare delivery system in New York City. More specifically, the authors discuss how education, training, and exercises, in combination with a unique emergency management approach, prepared the city to respond to several infectious disease threats and outbreaks including Ebola, Zika, seasonal influenza, drug-resistant fungus, and a 2018 measles epidemic—the largest measles outbreak in the United States in decades.

Part One concludes with Chapter 3 and a discussion of another response to a measles outbreak—this one occurring in 2019 in Clark County, Washington, USA. Clark County has a population of nearly 500,000 and is located in southwest Washington—across the Columbia River from Portland, Oregon. In this chapter, the authors discuss how local health officials in Clark County, Washington, and Oregon collaborated to implement disease surveillance, social distancing measures, and media monitoring. They go on to discuss the challenges related to managing an influx of resources resulting from emergency declarations and using a national incident command system during the response.

Chapter 4 begins the section of the book dedicated to methods that cities are using to address infectious disease risks. This chapter features a discussion of how after-action reviews (AARs)—a qualitative review of actions taken to respond to an event—have served as a tool for evaluating the response to urban disease outbreaks in Nigeria. AARs are an important tool that can inform best practices and help identify challenges and gaps in urban pandemic preparedness. The AARs conducted by the Nigeria Centre for Disease Control determined how actions were implemented (as opposed to how they were planned), identified processes that worked well, determined root causes for success and failure, and provided practical recommendations to improving preparedness in cities.

In Chapter 5, the authors discuss the challenges faced by city leadership in Blantyre, Malawi, in coordinating with other stakeholders and implement targeted HIV prevention programs and rapidly respond to HIV outbreaks. They then highlight how various stakeholders are working to develop of a highly specific, comprehensive strategy to optimize and institutionalize HIV prevention efforts and enable city and district leadership to sustainably reduce new HIV infections, as well as how the approach can be scaled to other health-related conditions.

Chapter 6 includes a discussion on how the Kawasaki City, Japan, has built a robust and advanced early detection and response system that has become a state-of-the-art model for infectious disease control in urban areas in Japan. They also emphasize the importance and impact of multisectoral engagement and how it has been fostered through joint exercises between public health authorities, medical institutions, security sector, and private sector actors.

Chapter 7 features a discussion of preparedness in frontline hospitals in the city of Phoenix. This chapter describes in detail vulnerabilities of urban hospitals in the United States, the tiered system used for the response to priority pathogens, maintaining preparedness, and how a six-hospital system in Phoenix has addressed several of these unique threats through a gap analysis, the development of subcommittees, the provision of enhanced personal protective equipment kits, electronic medical records, engineering controls, improved communication, and drills and training.

Part Two concludes with Chapter 8 discussing the development of emergency operation centers in Myanmar and how they contribute to pandemic preparedness in cities. In 2019, numerous infectious disease outbreaks occurred in Myanmar cities, including seasonal influenza, H1N1, H3N1, Chikungunya, and dengue, which prompted the Ministry of Health and Sports to fast-track the establishment of public health emergency operations centers in order to better respond to urban pandemics. This chapter highlights policy challenges and technical and operation-related issues in developing and strengthening the public health emergency operations centers for urban pandemic preparedness—and how they were overcome—in a limited resource setting that is undergoing rapid urbanization.

Chapter 9 features a discussion on an urban outbreak simulation and highlights how such exercises can create value and inform humanitarian response. This chapter further emphasizes key analytic takeaways and insights borne through these efforts and identifies opportunities for work to promote preparedness in cities for real infectious disease threats, such as COVID-19.

Chapter 10 focuses on the relationship between a city's population, its form, structure and function, and the environment in which the population lives and interacts. The authors explore these ideas through the context of a major epidemic threat posed by smallpox. Although this disease was eradicated in 1980, the pathogen historically accounted for large epidemics with high mortality, and recent advances in synthetic biology and genetic engineering make the accidental or malicious re-emergence of smallpox possible. In this chapter, Sydney, Australia, and Phoenix, USA, are used as case studies to illustrate how social, economic, health, and other factors are important determinants of predicted outcomes in these cities were they subject to a smallpox outbreak.

Finally, while governments have a recognized responsibility for the health security of their citizens, the continuity of urban social structures and activities during a crisis will be significantly affected by the response of private organizations. In Chapter 11, authors from International SOS discuss the role of the private sector in urban health security and present case studies of transport, retail, and hospitality industry planning and response. This chapter also presents corporate best practices for global health security planning and response and how those practices can improve urban preparedness.

In summation, this collection of chapters presents an overview of the experiences of cities in responding to infectious disease outbreaks and how cities are preparing for and responding to some of the largest existential issues they face.

Matthew Boyce and Rebecca Katz

References

[1] Armelagos GJ, Goodman AH, Jacobs KH. The origins of agriculture: population growth during a period of declining health. Popul Environ 1991;13:9−22.

[2] Balter M. The goddess and the bull: Catalhöyük - an archaeological journey to the dawn of civilization. New York: Routledge; 2016.

[3] Uruk Project. The City of Uruk, 3000 B.C.: short historical reference. 2013. staff.scm.uws. edu.au/~anton/Research/Uruk_Project/History.html [Accessed 14 December 2019].

[4] Reiff JL, Keating AD, Grossman JR, editors. The Electronic encyclopedia of Chicago. Chicago: Chicago Historical Society; 2005.

[5] Agwanda A, Amani H. Population growth, structure and momentum in Tanzania. Dar es Salaam: The Economic and Social Research Foundation; 2014.

[6] Zhang HX. Rural livelihoods in China: political economy in transition. New York: Routledge; 2015.

[7] Guangzhou Statistics Bureau. Statistical yearbook. 2010. data.gzstats.gov.cn/gzStat1/chaxun/njsj.jsp [Accessed 14 Dec 2019].

[8] United Nations Human Settlements Programme. World cities report. New York: United Nations; 2016.

[9] United Nations Department of Economic and Social Affairs. World urbanization prospects: the 2018 revision. New York: United Nations; 2018.

[10] Boyce MR, Osman R, Katz R. A research agenda for urban pandemic preparedness: a call to action. Washington: Center for Global Health Science & Security; 2019.

[11] Phillips DR. Urbanization and human health. Parasitology 1993;106:93−107.

[12] Papagrigorakis MJ, Yapijakis C, Synodinos PN. Typhoid fever epidemic in ancient athens. In: Raoult D, Drancourt M, editors. Paleomicrobiology: past human infections. Berlin: Springer; 2008.

[13] Cherry JD. The chronology of the 2002−2003 SARS mini pandemic. Paediat Resp Rev 2004;5:262−9.

[14] Alirol E, Getaz L, Stoll B, Chappuis F, Loutan L. Urbanisation and infectious diseases in a globalised world. Lancet Infect Dis 2011;11:131−41.

[15] Glaeser E. The triumph of the city: how our greatest invention makes us richer, smarter, greener, healthier, and happier. New York: Penguin Press; 2011.

[16] Morse SS, Mazet JA, Woolhouse M, Parrish CR, Carroll D, Karesh WB, et al. Prediction and prevention of the next pandemic zoonosis. Lancet 2012;380:1956−65.

[17] Black FL. Measles endemicity in insular populations: critical community size and its evolutionary implication. J Theoret Biol 1966;11:207−11.

[18] Barnes I, Duda A, Pybus OG, Thomas MG. Ancient urbanization predicts genetic resistance to tuberculosis. Evolution 2011;65:842−8.

[19] Gonçalves B, Balcan D, Vespignani A. Human mobility and the worldwide impact of intentional localized highlypathogenic virus release. Scient Rep 2013;3:810.

[20] Hertzberg VS, Weiss H. On the 2-row rule for infectious disease transmission on aircraft. Ann Glob Health 2016;82:819−23.

[21] Gómez JM, Verdú M. Network theory may explain the vulnerability of medieval human settlements to the Black Death pandemic. Scient Rep 2017;7:43467.

[22] Volz EM, Miller JC, Galvani A, Ancel Meyers L. Effects of heterogeneous and clustered contact patterns on infectious disease dynamics. PLoS Comput Biol 2011;7:e1002042.

[23] Fournié G, Guitian J, Desvaux S, Cuong VC, Dung DH, Pfeiffer DU, et al. Interventions for avian influenza A (H5N1) risk management in live birdmarket networks. Proc Natl Acad Sci USA 2013;110:9177—82.

[24] Han J, Jin M, Zhang P, Liu J, Wang L, Wen D, et al. Epidemiological link between exposure to poultry and all influenza A(H7N9) confirmed cases in Huzhou city, China, March to May 2013. Euro Surveill 2013;18:20481.

[25] Gilbert M, Golding N, Zhou H, Wint GRW, Robinson TP, Tatem AJ, et al. Predicting the risk of avian influenza A H7N9 infection in live-poultry marketsacross Asia. Nat Commun 2014;5:4116.

[26] Karesh WB, Dobson A, Lloyd-Smith JO, Lubroth J, Dixon MA, Bennett M, et al. Ecology of zoonoses: natural and unnatural histories. Lancet 2012;380:1936—45.

[27] Vlahov D, Freudenberg N, Proietti F, Ompad D, Quinn A, Nandi V, et al. Urban as a determinant of health. J Urban Health 2007;84:16—26.

[28] Ompad DC, Galea S, Caiaffa WT, Vlahov D. Social determinants of the health of urban populations: methodologic considerations. J Urban Health 2007;84:42—53.

[29] Lai YJ, Lai HH, Chen YY, Ko MC, Chen CC, Chuang PH, et al. Low socio-economic status associated with increased risk of dengue haemorrhagic fever in Taiwanese patients with dengue fever: a population-based cohort study. Trans Royal Soc Trop Med Hyg 2020;114:115—20.

[30] Chen WK, Wu HD, Lin CC, Cheng YC. Emergency department response to SARS, Taiwan. Emerg Infect Dis 2005;11:1067—73.

[31] Patel RB, Burke TF. Urbanization—an emerging humanitarian disaster. New England J Med 2009;361:741—3.

[32] Neiderud CJ. How urbanization affects the epidemiology of emerging infectious diseases. Infect Ecol Epidemiol 2015;5:27060.

[33] Armelagos GJ, Barnes KC, Lin J. Disease in human evolution: the re-emergence of infectious disease in the third epidemiological transition. Natl Mus Natl Hist Bull Teach 1996;18:1—6.

[34] Hotez PJ. Global urbanization and the neglected tropical diseases. PLoS Negl Trop Dis 2017;11:e0005308.

[35] National Academies of Sciences. Engineering, and Medicine. Urbanization and slums — infectious diseases in the built environment: proceedings of a workshop. Washington: The National Academies Press; 2018.

[36] Archeoclub d'Italia Sede di Venezia Onlus. Lazzaretto vecchio. 2016. www.lazzarettov ecchio.it [Accessed 14 December 2019].

[37] Barber B. If mayors ruled the world: dysfunctional nations, rising cities. New Haven: Yale University Press; 2013.

[38] Rinchiuso-Hasselmann A, Starr DT, McKay RL, Medina E, Raphael M. Public compliance with mass prophylaxis guidance. Biosec Bioterror 2010;8:255—63.

[39] Hanvoravongchai P, Adisasmito W, Chau PN, Conseil A, de Sa J, Krumkamp R, et al. Pandemic influenza preparedness and health systems challenges in Asia: results from rapid analyses in 6 Asian countries. BMC Public Health 2010;10:322.

[40] Salinsky E. Governmental public health: an overview of state and local public health agencies. Washington: National Health Policy Forum; 2010.

[41] Agendas Kingdon J. Alternatives and public policies. Boston: Little Brown; 1984.

[42] McNeil Jr DG. Can the government require vaccinations? yes. New York Times; April 11, 2019.

[43]   Li Q, Guan X, Wu P, Wang X, Zhou L, Tong Y, et al. Early transmission dynamics in Wuhan, China, of novel coronavirus—infected pneumonia. New England J. Med. 2020;382:1199−207.

[44]   World Health Organization. Novel coronavirus (2019-nCoV) situation report − 11. Geneva: WHO; 2020.

[45]   World Health Organization. Coronavirus disease 2019 (COVID-19) situation report − 40. Geneva: WHO; 2020.

[46]   Ghebreyesus TA. WHO director-general's opening remarks at the media briefing on COVID-19. World Health Organization; March 11, 2020.

Chapter 1

Controlling dengue, an urban pandemic — a case study of Delhi, India

Shyamala Mani[1], Samayita Ghosh[1], Richa Sharma[1], Aditya Ajith[2], Poornima Prabhakaran[1]

[1]Centre for Environmental Health, Public Health Foundation of India, Gurgaon, Haryana, India;
[2]Chief Minister's Urban Leaders Fellow, Government of NCT of Delhi, New Delhi, India

Dengue virus (DENV) is a mosquito-borne pathogen that causes up to ~100 million dengue cases each year, globally, placing major public health, social, and economic burdens on numerous low and middle-income countries (LMICs) [1]. The infection is transmitted by female *Aedes aegypti* mosquitoes that acquire the infection from an infected person and are able to transmit the virus to another healthy person within one week. Traditionally, there are four dengue virus serotypes that have been associated with both mild (e.g., fever, rash) and more severe (e.g., hemorrhagic fever shock syndrome) symptoms, in addition to a range of neurological and other clinical complications, potentially with sequelae or fatal consequences. Dengue fever is usually characterized by an abrupt onset of febrile illness accompanied by frontal headache and retro-orbital pain, followed by a variety of possible clinical symptoms such as myalgia, arthralgia, vomiting, and nausea. The incubation period is between 3 and 8 days following the bite from an infected mosquito.

The World Health Organization recommends disease control strategies that capitalize on coordination and cooperation between intergovernmental agencies, government bodies, partners, and other organizations. According to its guidelines, dengue morbidity can be reduced by implementing improved outbreak prediction and detection through coordinated epidemiological and entomological surveillance; promoting the principles of integrated vector management; deploying locally-adapted vector control measures, including effective urban and household water management; and through communication to achieve behavioral outcomes in prevention programs. Further, dengue mortality can be reduced by implementing early case detection and referral

Inoculating Cities. https://doi.org/10.1016/B978-0-12-820204-3.00001-2

1

systems for patients; managing severe cases with appropriate treatment; reorienting health services to cope with dengue outbreaks; and training health personnel at all levels of the health system.

However, there is a growing consensus that no single intervention will be sufficient for controlling dengue disease. This is due to heterogeneities in the mosquito vector, viral pathogen, and human host factors that drive the complexity of transmission [2]. Given the global paradigms of vector and case management protocols, some countries such as Singapore and Thailand have used effective surveillance and prediction mechanisms to combat the disease at the early stages of outbreaks. This is achieved through the introduction of new tools such as the geographical information system (GIS) for the management of databases and spatial identification of "hotspots" and the use of hand-held terminals for the collection of field surveillance data. New initiatives also include regular mosquito surveillance of housing estates and crowded places, redesigning of structural habitats, incorporation of heating elements in roof gutters, and tracking behavioral changes.

Over the past decade, changes in the patterns of cases and deaths recorded across India have presented an opportunity to analyze the strategies and actions used by various states to control and manage the disease at the local-level. Given the frequent occurrence of concurrent infections with multiple DENV serotypes, the city of Delhi has established itself as a hyperendemic environment. This chapter seeks to explore the documented efforts of local actions in Delhi to manage DENV outbreaks.

A brief background on the dengue virus

Dengue virus serotypes and associated virulence

Dengue viruses are positive-stranded RNA viruses in the *Flavivirus* genus in the *Flaviviridae* family. As mentioned above, there are four distinct DENV serotypes that share antigenic relationships — DENV-1, DENV-2, DENV-3, and DENV-4. The DENV-2 and DENV-3 serotypes have been described as virulent genotypes and are commonly associated with more severe disease outcomes [3]. Several studies indicate that although the DENV-4 is associated with severe outcomes, it appears to be the most clinically mild [4,5]. DENV-2 and DENV-4 have been associated with increased disease severity as a secondary infection, whereas DENV-1 and DENV-3 seem to cause more severe disease in primary infection compared to the other two serotypes [6].

Although infection with one serotype confers lifelong protection against that serotype, it does not necessarily protect against secondary infection with a heterologous serotype [7]. A secondary DENV infection results when a person previously infected with one serotype is exposed to a different serotype, and it has been documented as the single most important risk factor for severe dengue [8]. In fact, non-protective but cross-reactive antibodies may enhance disease severity [9].

Diagnosis and treatment of dengue

Several methods can be used to diagnose DENV infections. These include virologic tests that directly detect elements of the virus (e.g., RT-PCR) and serological tests that detect human-derived immune components produced in response to the virus (e.g., ELISA). Depending on the time of patient presentation, the application of different diagnostic methods may be more or less appropriate. For instance, patient samples collected during the first week of illness should be tested by both serological and virologic methods. Serological methods can confirm the presence of a recent or past infection, with the detection of IgM and IgG anti-dengue antibodies. IgM antibodies are detectable approximately 1 week after infection and levels are highest at 2—4 weeks after the onset of illness. They remain detectable for about 3 months. The presence of IgM is indicative of a recent DENV infection. IgG antibody levels take longer to develop than IgM, but IgG remains in the body for years and the presence of IgG is indicative of a past infection [10].

No treatment currently exists for dengue. However, a vaccine — *Dengvaxia* — is licensed and available in some countries for people between 9 and 45 years of age. The World Health Organization recommends that the vaccine should only be given to persons with a confirmed DENV infection given that in 2017 the vaccine manufacturer, Sanofi Pasteur, announced that people who receive the vaccine and have not been previously infected with a dengue virus may be at risk of developing severe dengue if they contract dengue after being vaccinated [11].

Global distribution of the dengue virus

Globally, more than 3.5 billion people are at risk of infection with DENV and an estimated 390 million cases are detected annually. The global epidemiology is characterized by the simultaneous circulation of multiple serotypes and endemic dengue hemorrhagic fever/dengue shock syndrome in affected countries. Dengue viruses have evolved rapidly with its spread worldwide, and genotypes associated with increased virulence have expanded from South and Southeast Asia into the Pacific and the Americas. Over the past three decades, nearly 3 million children have been hospitalized with this disease, mainly in South-East Asia. Recent outbreaks in the Pacific Islands, China, India, Sri Lanka, Cuba, and Venezuela indicate the high intensity and rapid spread of dengue transmission which is closely connected with urbanization, as well as increases in temperature and precipitation and changes in spatial patterns [12,13].

Although Dengue is the most common arbovirus infection globally, with an incidence that has increased dramatically over the past 50 years, the burden is poorly quantified. One study conservatively estimates through cause of death ensemble modeling that there has been an average of 9221 dengue deaths per year between 1990 and 2013, increasing from a low of 8277 (95% uncertainty estimate: 5353–10,649) in 1992, to a peak of 11,302 (95% uncertainty estimate: 6790–13,722) in 2010 [14]. This yielded a total of 576,900 (330,000–701,200) years of life lost to premature mortality attributable to dengue in 2013. This study also estimated that the incidence of dengue more than doubled every decade, from 8.3 million (3.3 million–17.2 million) apparent cases in 1990, to 58.4 million (23.6 million–121.9 million) apparent cases in 2013. When accounting for disability from moderate and severe acute dengue, and post-dengue chronic fatigue, 566,000 (95% uncertainty estimate: 186,000–1,415,000) years lived with disability were attributable to dengue in 2013. Considering fatal and non-fatal outcomes together, dengue was responsible for 1.14 million (0.73 million–1.98 million) disability-adjusted life-years in 2013.

The historical and current distributions of dengue in India

Dengue is driven by a complex interaction between the host (i.e., primates, including humans), the vector (i.e., A. aegypti mosquitoes), and the virus. These interactions are also influenced by climatic factors that characterize the variability in the extrinsic incubation period. Over the years, dengue has transitioned into an urban epidemic in India – characterized by starting in urban centers before gradually expanding to surrounding peri-urban and rural areas. This is because urban areas are both densely populated and experience high temperature levels owing to the urban heat island effect, and hence become more favorable for A. aegypti mosquitoes. The vector thrives in areas with standing water, including puddles, water tanks, containers and old tires. The lack of reliable sanitation, irregular garbage collection, and water storage practices also contribute to the breeding and spread of the mosquitoes in most Indian cities.

Research has indicated that all four-dengue serotypes circulate in India, with DEN-2 and DEN-3 being the more commonly reported serotypes. Several studies have also reported the circulation of multiple serotypes, which is known to increase the probability of secondary infection, leading to a higher risk of severe dengue disease [7].

Since the 1990s, India has been experiencing frequent dengue epidemics and the disease has gradually become one of the leading causes of hospitalization. Since the 1990s, epidemics of dengue have become more frequent in many parts of India. From 1998 to 2009, 82,327 dengue cases (incidence: 6.34 cases per million population) were reported; however, during the period from 2010 to 2014, 213,607 cases (incidence: 34.81 cases per million population) of dengue fever were observed (Fig. 1.1). Thus, the number of dengue cases during that period of 5 years increased markedly, by a factor of approximately 2.6, with respect to the 1998–2009 period [15]. Further, surveillance data from the National Vector Borne Disease Control Programme (NVBDCP) show a total of 683,545 dengue cases and 2576 deaths reported in India during

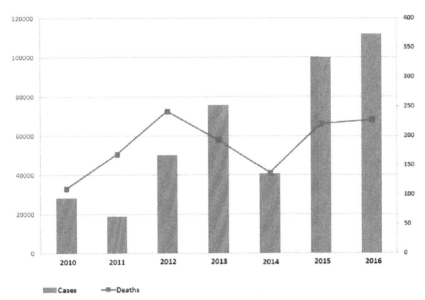

FIG. 1.1 Number of Dengue cases and deaths in India reported by the National Vector Borne Disease Control Programme from 2010 to 2016.

2009−17 (Fig. 1.2). These surveillance data also show an average of 28,227 dengue cases and 154 deaths reported annually from 2009 to 2012; the average number of cases reported increased thereafter, with an annual average of 100,690 cases from 2013 to 2017, but the reported number of deaths did not increase proportionately. Accordingly, India also experienced high dengue incidence in 2012 (41 cases per million population), 2013 (61 cases per million population) and 2014 (32 cases per million population). Information regarding the severity of dengue cases is not available from NVBDCP surveillance data.

On a more granular level, since 2010, the states of Assam, Bihar, Jharkhand, Orissa and Uttarakhand and some union territories including the Andaman and Nicobar Islands, Dadra and Nagar Haveli, and Daman and Diu have become endemic for dengue. State-level variability in incidence and mortality rates, as well as the circulation of multiple DENV serotypes necessitates a thorough inspection of local-level responses and case management. From 2014 to 2015, Delhi was the worst affected city, wherein cases increased from 995 to 15,827 cases, but other cities such as Bengaluru, Mumbai, and Kolkata have also been impacted by dengue outbreaks [17]. However, after this peak, Delhi has witnessed a steep decline in reported cases from 15,867 in 2015 to 4431 cases in 2016. This figure has remained relatively steady in the following years. Mortality in Delhi has also declined from 60 reported deaths in 2015 to 2 in 2019.

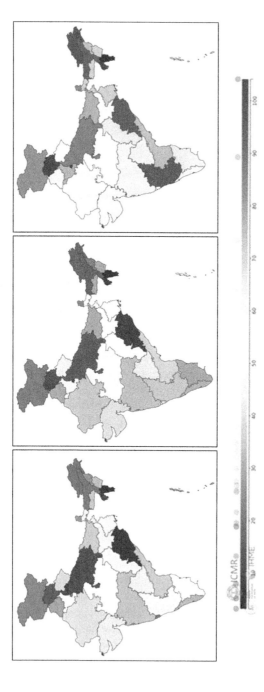

FIG. 1.2 Heat maps representing decadal shifts in disability-adjusted life years per 100,000 population during 1995, 2005 and 2015, for all age groups [16].

Studies indicate that most of the laboratory confirmed dengue cases in India occurred in young adults. Dengue positivity was higher between the months of August and November, corresponding to monsoon and post-monsoon season in most states in India. The computed pooled estimate of case fatality due to dengue in India is 2.6% with a high level of variability in reported case fatality rates [18].

Dengue control in the city of Delhi

Delhi and its urbanization

Delhi is a metropolitan city with a geographical area of 1484 km^2 (Fig. 1.3). Delhi is the most populous city in India with over 30 million people inhabiting the city [19], and is divided into nine districts: North, North West, West, South West, South, East, North East, Central, and New Delhi. Delhi experiences a typical humid subtropical climate with temperature ranging from 5 to 40 °C (41−104 °F) and an average annual rainfall of 714 mm (28 in). Most of the rainfall in the city occurs in the monsoonal months of July and August and the city experiences a mean relative humidity of around 66%.

There are five factors − demographic, social, agricultural, pathogenic, and infrastructural − that have been suggested for the surge of dengue epidemics [20].

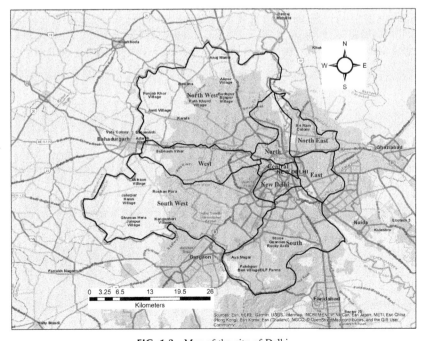

FIG. 1.3 Map of the city of Delhi.

All five have been observed in Delhi, rendering the city a potential site for an outbreak. For example, Delhi has witnessed a steep rise in population due to population growth as well as high migration rates in the city (Fig. 1.4). Such a massive and unprecedented rise in population has resulted in unplanned and uncontrolled urbanization in the city leading to infrastructural challenges. Furthermore, social factors including behavioral considerations such as water storage practices can impact the availability of breeding grounds for mosquitoes. Like many other developing nations, the basic safety nets of healthcare in India are also inadequate, although governments have been working toward improving the basic health infrastructure including surveillance and vector control. Additionally, local climate conditions, such as urban heat islands, also have the potential to impact dengue outbreaks in Delhi.

The warm and humid monsoon season in Delhi provides a favorable breeding environment for *Aedes* mosquitoes. Apart from climatic factors, other socio-economic factors that contribute to dengue having an epidemic potential in Delhi are a poor water management system, inadequate sanitation services, slums and informal settlements, and a lack of awareness [22].

Historical trends of dengue in Delhi

Delhi has been recording dengue cases every year since 1996 and the disease is hyper-endemic in the city [23]. A 1996 dengue outbreak in Delhi is perhaps the most infamous, when 10,252 cases and 253 deaths, resulting in a mortality rate of 4.1%, were reported. The 1996 dengue epidemic that occurred around Delhi, subsequently spread all over the country resulting in at least 16,000 cases and 545 deaths. The dengue incidence sharply increased from 1998 to

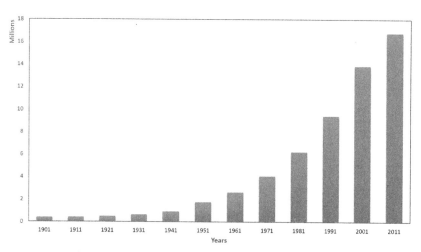

FIG. 1.4 Population growth in the city of Delhi, 1901–2011 [21].

2001 from 0.72 to 3.21 cases per million population. In 2003, 2005, 2006, 2008 and 2009, the incidence of dengue cases exceeded 10 per million population, and an incidence of greater than 15 cases per million population has been reported annually since 2015.

While all four serotypes co-circulate in the city, Delhi faces the threat of dengue outbreaks every 3 or 4 years exhibiting variations in serotype dominance. From 2003 to 2006, DENV-3 dominated the dengue outbreaks in Delhi, 2007−09 was dominated by DENV-1, and DENV-2 dominated from 2011 to 2014 [24,25]. DENV-2 was also responsible for the dengue outbreaks in 1996 [26]. Between a ten-year period, from 2001 to 2011, the city reported 20,289 cases and witnessed 140 deaths (0.69% case fatality rate). This decade included three distinct dengue epidemics with between 2800 and 6200 cases reported annually (Fig. 1.5). Even during non-epidemic years, between 45 and 1216 cases of dengue were reported in Delhi. The most recent outbreak in the city, which occurred in 2015, included 15,875 reported cases and 60 deaths [16]. Of importance, the past decade has also seen time between major municipal outbreaks decrease, as they have occurred in 2006, 2010, 2013 and 2015.

Spatial distribution of dengue in Delhi

Spatial data for Delhi from the Census of India and local authorities is available at aggregated levels including wards or districts. However, to assess risk factors at the local-level, additional data need to be collected. Colonies are administrative spatial units that are used primarily for property tax collection and investments, but this level is also appropriate for assessing dengue risk factors. Integrating various data from a variety of sources (e.g., land use, population density, property tax ranking, etc.) for each colony in a GIS environment can aid in understanding local socio-environmental heterogeneity.

Although difficult, some have attempt to do this. Research analyzing the spatio-temporal distribution of dengue cases and the extent of clustering based on data from the Delhi surveillance system from 2008 to 2010 ultimately found that increasing distance from the forest in Delhi reduced the risk of dengue infection [28]. The study also demonstrated high heterogeneity in incidence rates within areas with the same socio-economical profiles along with considerable inter-annual variability. Research has also investigated clustering and found that the dengue case cluster size is dependent upon its emergence time. In 2008, the highest density of cases was located in West, Central, and East Delhi (Fig. 1.6) [28]. Overlapping epidemiological data with environmental typology information, researchers found the density of dengue cases was highest in areas characterized by poor access to urban infrastructure and a high population density. Subsequent work in 2009 found that Central, East Delhi, and South Delhi, were impacted; while in 2010, North-West,

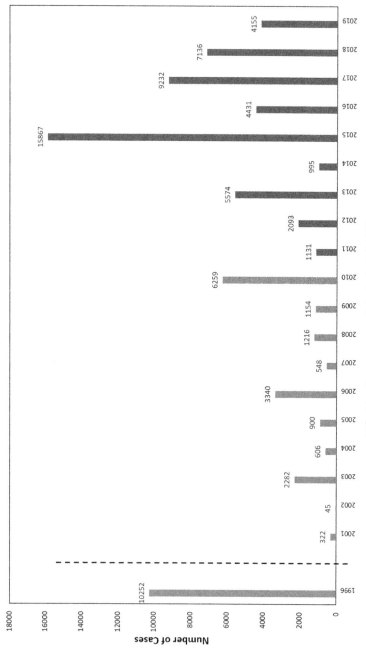

FIG. 1.5 Dengue cases and deaths in Delhi [27].

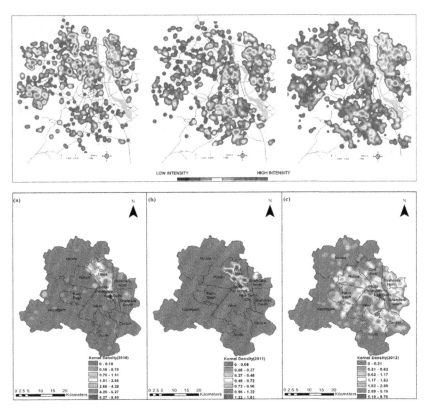

FIG. 1.6 Kernel density estimation (KDE) of individuals registered in the sentinel hospitals of Delhi (2008, 2009, 2010) [28]; KDE for dengue incidences in Delhi (2010, 2011, 2012) [29].

Central, and East Delhi were greatly impacted. South Delhi and North Delhi also show a relatively higher number of cases in comparison to 2008 and 2009. Others have found that the infection density in 2010 and 2011 was highest for New Delhi and Civil Lines zones [29]. While in 2012, the density was highest for Sadar Paharganj, New Delhi, Shahadra North and Shahadra South neighborhoods, as well as in parts of the Rohini, Civil Lines, Karol Bagh neighborhoods, and north eastern parts of South and Central districts.

Other work has shown that the Central Delhi, Civil Lines, Rohini neighborhoods, and the South and West Districts have the highest odds of dengue occurrences [30]. Household surveys in Delhi among 3350 households in low income neighborhoods found a very high number of confirmed dengue cases in the surveyed wards in East Delhi and North Delhi [31]. Still, the high number of cases may be the result of more mosquitoes. This hypothesis is supported by work that has found that low income group localities witnessed the highest breeding of *Aedes* mosquitoes, with plastic containers being the primary containers used by mosquitos for breeding [31].

Delhi's actions for controlling dengue

Given the diversity of urban systems and the variability of response structures in place, India continues to prioritize the intensification of the interface between systems and affected communities through manual notification processes to ensure context-driven case management. In India, case detection, case management, and vector control are the main strategies for prevention and control of dengue virus transmission and Delhi's response actions are understood within this larger context. Therefore, a majority of the city's documented efforts are part of larger, national protocols. But while most of Delhi's actions are in compliance with the national guidelines, the city has focused extensively on the implementation efforts especially by mobilizing community action with an emphasis on preventive strategies. The government has also strengthened primary and community health centers that can be used for dengue surveillance, management, and response. Strong political will coupled with community-action has worked effectively in controlling dengue in Delhi. The observed reductions in dengue cases and mortality and the improvement of recovery rates are indictors of Delhi's success in managing the disease by leveraging existing systems, strengthening primary healthcare infrastructure, focused communication, and participatory action.

In response to the cyclic dengue outbreaks, Delhi developed and implemented a municipal surveillance system that, as of 2010, consisted of 33 public hospitals and three private hospitals. Such work is important for monitoring trends in dengue distribution, seasonal patterns and circulating serotypes to guide dengue control activities with efforts currently being made to ensure early warning signals for timely detection of outbreaks [22]. The decision to form a dedicated dengue control cell to conduct prevention-related activities across the city has also helped to identify high-risk areas. This information was then shared with local bodies so that they could take necessary preventive measures. All establishments, including government and private hospitals, were also directed to identify a nodal officer who would work with the dengue control cell to ensure that there was no mosquito breeding on the premises.

But apart from monitoring spatial distribution of the disease and surveillance, there are other various efforts focused on proactive vector control that decentralize action and the Delhi government, in collaboration with the NVBDCP, has put forward measures to escalate action and curb dengue outbreaks. For instance, under the recommendation of the Ministry of Health and Family Welfare, Government of India, the city celebrates National Dengue Day on May 16th of each year and works to increase awareness about dengue and to intensify preventive measures and preparedness for the control of the disease before transmission season starts. The city has also intensified media engagement as well as community-based communication toward the fulfillment of the objective.

Other efforts to generate community action include a call-to-action titled "10 Hafte, 10 Baje, 10 Minute − Har Ravivar, Dengue Par Vaar," which translates to "10 Weeks, 10 O'clock, 10 Minutes − Each Sunday, Anti-Dengue measures." Issued by the Delhi government, this initiative encourages citizens to spare 10 min each Sunday morning for ten weeks during the dengue season to check for mosquito breeding, drain and change water in containers, treat areas of water accumulation that cannot be drained with oil/kerosene, and encourage emptying clear standing water. These efforts were supported by the health and sanitary departments, hospitals, resident welfare associations, and the general. As part of the community mobilization efforts, citizens are also encouraged to take a selfie of their inspection, and share it on a government-provided WhatsApp number.

The city has also held training and awareness campaigns with various stakeholders such as resident welfare associations, schools, and others. These campaigns generally seek to increase awareness about the disease and public participation in efforts leading to improved disease control.

The municipality has also taken measures to improve the treatment of dengue. For instance, the Delhi government has made exemplary efforts at strengthening the primary healthcare system through the development of the Mohalla Clinic model (i.e., neighborhood clinic model) that provides diagnostic and out-patient services, 109 essential medicines, and over 200 medical tests, including those for mosquito-borne diseases. The government has also issued orders capping the rates of medical services provided. Between 2015 and 2016, the government built 106 such clinics that have served more than 16.24 million residents [32]. Additionally, all government hospitals (33), dispensaries (262) and local clinics (106) have dedicated fever corners with sensitized and trained medical and paramedical staff regarding the prevention and management of dengue. To this end, 1000 beds have been earmarked for fever cases in Delhi's government hospitals. The government also issued circular relaxing norms for private hospitals to increase the bed strength for treating dengue patients − allowing private hospitals and nursing homes to increase their bed strength by 10%−20% during dengue season to accommodate febrile patients. This is done under the provisions of the regulatory framework that governs the private healthcare system in India (Fig. 1.7).

Challenges and next steps for controlling dengue in Delhi

While the government has taken many measures to control the dengue within the city and minimize the epidemic, more work is needed and there is much room for improvement − particularly with regard to preventive actions, reactive measures, and improved knowledge and awareness.

DELHI GOVERNMENT
PUBLIC HEALTH INFRASTRUCTURE

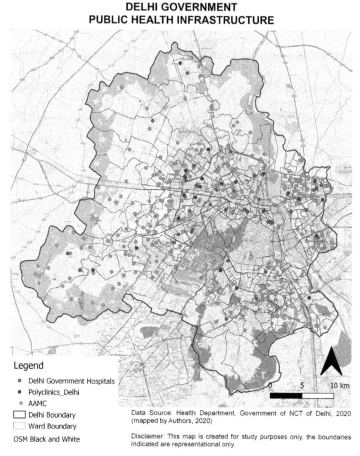

FIG. 1.7 Delhi's public health infrastructure.

Preventive actions

While Delhi has implemented surveillance strategies and taken action to better understand the local burden of dengue — which are necessary for making decisions regarding appropriate prevention and control strategies — the city could borrow strategies from other cities in India. The city of Bangalore represents one potential example. Bangalore has supplemented national protocols and intensified data collection by implementing a hospital-based surveillance system that maps both dengue risk factors and spatial determinants by making home-visits to dengue-positive patients to inspect the surrounding areas. The patient's habitation information, along with their immunization record and other relevant information, is then entered into a database and monitored.

Furthermore, the risk of dengue is now known to be highly dependent on climatic factors — especially precipitation. For example, there is a lag phase between rains and the appearance of dengue cases in Delhi. Lags average approximately two months, and while most efforts in Delhi currently rely on a reactive approach, these lags provide a unique window of opportunity for implementing public health interventions, such as carrying out proactive vector control measures [33]. This lag also provides enough time for resource mobilization to ensure the effective implementation of medical measures to minimize the health impacts of dengue outbreaks.

Additionally, with regard to preventive measures, even during the non-transmission seasons, environmental surveys should be conducted — especially in areas with poor environmental sanitation, poor housing, and inadequate water supplies. Work has demonstrated that two types of breeding foci exist for the dengue-carrying *Aedes* mosquitoes — primary and secondary [34]. Primary foci, such as overhead tanks and curing tanks, usually serve as mosquito breeding grounds throughout the year and are used for breeding during the non-transmission season. However, during the monsoonal season, mosquito larvae spread to the secondary foci, such as plastic containers, discarded tires, and desert coolers [21,35]. Hence, the city could improve vector control and prevent subsequent disease by working to locate such vulnerable environments and conduct more vigilant and regular monitoring of these areas.

A need also exists to develop weather-based forecasting models to provide policymakers and program managers with information well in advance to be prepared and take preventive measures. Early warning and weather-based forecasting systems are increasingly recognized as crucial components of resilience building, which could help prepare for other municipal emergencies as well. Such systems require improved environmental monitoring and health surveillance and depend on high-quality, long-term data on environment and climate related health outcomes. These data, unfortunately, do not currently exist. Accordingly, efforts to collect them could be helpful in understanding the relationship between dengue (and other vector-borne diseases) and the environment, as well as in forecasting disease outbreaks [36].

Reactive measures

Currently, Delhi relies heavily upon vector control strategies for controlling dengue. However, studies indicate that the *Aedes* mosquitos around Delhi are developing resistance to the insecticides that are currently administered to control vector populations [37]. Thus, there is a compelling need to conduct regular testing for insecticide susceptibility to inform prevention and control efforts and to manage resistance to insecticides.

City authorities also need to better understand the spatio-temporal distribution and clustering of dengue cases to identify high-risk and vulnerable areas. This could be done with smart solutions including command and control centers with real-time monitoring of dengue cases, or with more community focused efforts, such as citizen science. The latter offers multiple advantages including improved data collection that could facilitate real-time monitoring and research, as well as increased public awareness. Free or low-cost software could also be explored to facilitate decision support system approaches for the prevention and control of vector-borne diseases such as dengue in resource-poor environments. For instance, combining Google Earth and free GIS software (e.g. HealthMapper developed by the World Health Organization, and SIGEpi developed by the Pan American Health Organization) [38].

Knowledge and awareness

An important component of dengue management that necessitates greater focus in Delhi is knowledge and awareness raising, supplemented with effective communication. Improved knowledge about the vector, disease transmission, and the relationship between dengue and environmental factors could improve engagement with the city's population. To this end, the measures that are currently being implemented need to be evaluated to understand gaps and barriers, strategies need to be updated by including feedback, and successful practices and initiatives should be documented and communicated. This could serve as an opportunity to acquire knowledge and communicate lessons learned for improving and adapting strategies and activities for more sustained and effective interventions.

Further, while Delhi has carried out extensive awareness campaigns for the prevention of dengue these could be improved through subtle strategy shifts. For instance, targeting housewives could improve the impact of these activities, as these women are generally more engaged with water storage and cleanliness of the house and neighborhood. Governments have explored word of mouth and mass media, such as television and radio, as a means of communication and awareness generation. Still, these have not led to sufficient action − at least not for mosquito breeding control and the prevention of dengue. One approach for overcoming this could be door-to-door information dissemination conducted by health and municipality workers. Such efforts could include demonstrations on how to handle *Aedes* mosquito breeding sites including emptying water containers regularly. Applying repellents and using protective clothes during the day represent other behavioral modifications that should be promoted, as they can have a major, cumulative effect at the population level.

Conclusions

Delhi has successfully demonstrated a steady decline in the number of cases and deaths from 15,867 cases and 60 deaths in 2015 to 2036 cases and 2 deaths in 2019. An enhanced primary health infrastructure that improves access to low-cost health care services and a strategic course of citizen action have been integral to these improvements. Given that vector-borne diseases mostly affect populations residing in the city's fringes, pro-poor policies, welfare services, and participatory governance hold the potential to generate significant results in the management of dengue. The Mohalla Clinic is an exemplar of inclusivity wherein, it is principled around the provision of health care services not only to the permanent residents of the city but also migratory and floating population from neighboring states that constitute a considerable patient load.

However, there is work to be done and a recognized need for monitoring and measuring the effectiveness of dengue control strategies. Such work can identify gaps in planning and implementation and help Delhi better proactively prepare for dengue outbreaks, respond to outbreaks efficiently, and improve local-level knowledge and awareness about the disease.

References

[1] Katzelnick LC, Coloma J, Harris E. Dengue: knowledge gaps, unmet needs, and research priorities. Lancet Infect Dis 2017;17:e88−100.

[2] Achee NL, Gould F, Perkins TA, Reiner Jr RC, Morrison AC, Ritchie SA, et al. A critical assessment of vector control for dengue prevention. PLoS Neglected Trop Dis 2015;9:e0003655.

[3] Rico-Hesse R. Dengue virus virulence and transmission determinants. In: Rothman A, editor. Dengue virus. Current topics in microbiology and immunology. Berlin: Springer; 2010.

[4] Nishiura H, Halstead SB. Natural history of dengue virus (DENV)−1 and DENV−4 infections: reanalysis of classic studies. J Infect Dis 2007;195:1007−13.

[5] Fried JR, Gibbons RV, Kalayanarooj S, Thomas SJ, Srikiatkhachorn A, Yoon I-K, et al. Serotype-specific differences in the risk of dengue hemorrhagic fever: an analysis of data collected in Bangkok, Thailand from 1994 to 2006. PLoS Neglected Trop Dis 2010;4:e617.

[6] Van Kleef E, Bambrick H, Hales S. The geographic distribution of dengue fever and the potential influence of global climate change. TropIKA.net 2010. In press.

[7] Halstead SB. Biologic evidence required for Zika Disease enhancement by dengue antibodies. Emerg Infect Dis 2017;23:569−73.

[8] Changal KH, Raina A, Raina M, Bashir R, Latief M, Mir T, et al. Differentiating secondary from primary dengue using IgG to IgM ratio in early dengue: an observational hospital based clinico-serological study from North India. BMC Infect Dis 2016;16:715−22.

[9] Rodriguez-Roche R, Gould EA. Understanding the dengue viruses and progress towards their control. BioMed Res Int 2013;13.

[10] World Health Organization. Laboratory diagnosis and diagnostic tests. In: Dengue: guidelines for diagnosis, treatment, prevention and control. Geneva: World Health Organization; 2009.

[11] World Health Organization. Dengue vaccine: WHO position paper − July 2016. Wkly Epidemiol Rec 2016;30:349−64.

[12] Halstead SB. The XXth century dengue pandemic: need for surveillance and research. World Health Stat Q 1992;45:292−8.

[13] Messina JP, Brady OJ, Scott TW, Zou C, Pigott DM, Duda KA, et al. Global spread of dengue virus types: mapping the 70 year history. Trends Microbiol 2014;22:138−46.

[14] Stanaway JD, Shepard DS, Undurraga EA, Halasa YA, Coffeng LE, Brady OJ, et al. The global burden of dengue: an analysis from the Global Burden of Disease Study 2013. Lancet Infect Dis 2016;16:712−23.

[15] Mutheneni SR, Morse AP, Caminade C, Upadhyayula SM. Dengue burden in India: recent trends and importance of climatic parameters. Emerg Microb Infect 2017;6:1−10.

[16] Institute for Health Metrics and Evaluation. Global burden of disease India compare. 2017. https://vizhub.healthdata.org/gbd-compare/india. [Accessed 27 October 2020].

[17] Ministry of Health & Family Welfare, Government of India. National vector borne disease control Programme. 2019. https://nvbdcp.gov.in/. [Accessed 27 October 2020].

[18] Ganeshkumar P, Murhekar MV, Poornima V, Saravanakumar V, Sukumaran K, Anandaselvasankar A, et al. Dengue infection in India: a systematic review and meta-analysis. PLoS Neglected Trop Dis 2018;12:e0006618.

[19] Population Stat. World statistical data. 2020. www.populationstat.com. [Accessed 9 November 2020].

[20] Gubler DJ. Epidemic dengue/dengue hemorrhagic fever as a public health, social and economic problem in the 21st century. Trends Microbiol 2002;10:100−3.

[21] Census of India 2011, Government of India.

[22] Vikram K, Nagpal BN, Pande V, Srivastava A, Gupta SK, Anushrita VP, et al. Comparison of *Ae. Aegypti* breeding in localities of different socio-economic groups of Delhi, India. Int J Mosq Res 2015;18:20.

[23] Savargaonkar D, Sinha S, Srivastava B, Nagpal BN, Sinha A, Shamim A, et al. An epidemiological study of dengue and its coinfections in Delhi. Int J Infect Dis 2018;74:41−6.

[24] Afreen N, Naqvi IH, Broor S, Ahmed A, Parveen S. Phylogenetic and molecular clock analysis of dengue serotype 1 and 3 from New Delhi, India. PLoS One 2015;10:e0141628.

[25] Shrivastava S, Tiraki D, Diwan A, Lalwani SK, Modak M, Mishra AC, et al. Co-circulation of all the four dengue virus serotypes and detection of a novel clade of DENV-4 (genotype I) virus in Pune, India during 2016 season. PLoS One 2018;13:e0192672.

[26] Dar L, Broor S, Sengupta S, Xess I, Seth P. The first major outbreak of dengue hemorrhagic fever in Delhi, India. Emerg Infect Dis 1999;5:589.

[27] Central Bureau of Health Intelligence, Directorate general of health services, Government of India.

[28] Telle O, Vaguet A, Yadav NK, Lefebvre B, Daudé E, Paul RE, et al. The spread of dengue in an endemic urban milieu−the case of Delhi, India. PLoS One 2016;11:e0146539.

[29] Mala S, Jat MK. Geographic information system based spatio-temporal dengue fever cluster analysis and mapping. Egypt J Remote Sens Space Sci 2019;22:297−304.

[30] Agarwal N, Koti SR, Saran S, Kumar AS. Data mining techniques for predicting dengue outbreak in geospatial domain using weather parameters for New Delhi, India. Curr Sci 2018;114:2281−91.

[31] Daudé E, Mazumdar S, Solanki V. Widespread fear of dengue transmission but poor practices of dengue prevention: a study in the slums of Delhi, India. PLoS One 2017;12:e0171543.

[32] Lahariya C. Mohalla clinics of Delhi, India: could these become platform to strengthen primary healthcare? J Fam Med Prim Care 2017;6:1−10.

[33] Bisht B, Kumari R, Nagpal BN, Singh H, Kumar S, Gupta AK, et al. Influence of environmental factors on dengue fever in Delhi. Int J Mosq Res 2019;6:11−8.

[34] Katyal R, Gill KS, Kumar K. Seasonal variations in Aedes aegypti population in Delhi, India. Dengue Bull 1996;20:78−81.

[35]   Nagpal BN, Gupta SK, Shamim A, Vikram K, Srivastava A, Tuli NR, et al. Control of *Aedes Aegypti* breeding: a novel intervention for prevention and control of dengue in an endemic zone of Delhi, India. PLoS One 2016;11:e0166768.

[36]   Bush KF, Luber G, Kotha SR, Dhaliwal RS, Kapil V, Pascual M, et al. Impacts of climate change on public health in India: future research directions. Environ Health Perspect 2011;119:765−70.

[37]   Samal RR, Kumar S. Susceptibility status of *Aedes aegypti* L. against different classes of insecticides in New Delhi, India to formulate mosquito control strategy in fields. Open Parasitol J 2018;6:52−62.

[38]   Lozano-Fuentes S, Elizondo-Quiroga D, Farfan-Ale JA, Loroño-Pino MA, Garcia-Rejon J, Gomez-Carro S, et al. Use of Google Earth to strengthen public health capacity and facilitate management of vector-borne diseases in resource-poor environments. Bull World Health Organ 2018;86:718−25.

Chapter 2

Municipal healthcare delivery special pathogens preparedness and response in the city that never sleeps: the NYC Health + Hospitals' emergency management approach to infectious disease threats

Syra Madad[1], Priya Dhagat[1], Nicholas Cagliuso[2]
[1]Emergency Management, System-wide Special Pathogens Program, NYC Health + Hospitals, New York, NY, United States; [2]Emergency Management, NYC Health + Hospitals, New York, NY, United States

New York City (NYC) is often referred to as the city that never sleeps. With more than 8.6 million daily residents and visitors hailing from countries throughout the world, speaking 800 languages, the "Big Apple" is a global social, political, and economic nexus — a living laboratory for what is possible, both fresh opportunities and significant threats. This chapter describes the unique challenges and innovative strategies used by NYC Health + Hospitals, the largest municipal healthcare delivery system in the United States of America, to better prepare for and respond to a multitude of infectious disease threats — everything from common infections, to emerging infectious diseases, to high-consequence pathogens.

The 2018–19 New York City measles outbreak was the largest such outbreak in the United States since 1992. Faced with the threat of a highly infectious disease with a 90% attack rate among susceptible persons coupled with a large anti-vaccine movement, the resurgence of measles in a large urban population posed numerous challenges to public health and healthcare delivery

Inoculating Cities. https://doi.org/10.1016/B978-0-12-820204-3.00002-4

systems. To respond to the outbreak, NYC Health + Hospitals utilized an emergency management approach to rapidly identify and isolate patients, increase infection control measures, monitor medical facility exposures, address vaccine delays, combat anti-vaccination propaganda, communicate risk effectively, and implement various administrative, engineering and system controls.

Additionally, the chapter offers experience-based perspectives on NYC Health + Hospitals' first-of-its-kind System-wide Special Pathogens Program's all-infectious disease hazards approach. This approach has been used in a variety of situations — ranging from supporting the system's safe and successful response in treating NYC's single confirmed-Ebola patient, to aiding in the City's responses to Zika virus, seasonal influenza, and drug-resistant *Candida auris*. The chapter also discusses the development and implementation of a frontline special pathogens course, all within the rubric of the Program's emergency management-centric mission to care for all, regardless of ability to pay or immigration status and rooted in the premise that, "ready or not patients will present."

NYC urban landscape and constant threat of infectious diseases of public health concern

As the most populous city in America with two major international airports and over 100 million travelers annually, New York City is home to 8.3 million residents with a population density of 26,403 people per square mile [1]. Such an extreme population density can be a catalyst for the rapid spread of infectious diseases. Many various risk factors in the urban environment can contribute to the proliferation of infectious diseases including environmental, social, and political drivers (Table 2.1). Cities like NYC are at risk given the ever-changing drivers that influence emerging and re-emerging infectious diseases.

Cities, as Jane Jacobs notes, have the capability of providing something for everybody, only because, and only when, they are created by everybody [2]. This embodies the implicitly collaborative nature of cities, where individuality, at its most elemental level, rises to scale and helps define the essence of our experiences with them. More recently, Beauregard argues that cities are cauldrons for four contradictions [3]. First, they are defined by their wealth and poverty; second, cities are, at once, environmentally damaging with the promise of sustainability; third, complex political apparatuses govern these urban environs; and fourth, city living often advances tolerance among disparate groups.

In the context of managing infectious diseases, this all weighs heavily, as wealth and poverty are strong determinants of both access to healthcare services and health status. Even with their efforts to steer toward sustainability, cities' untoward environmental impacts are self-evident and further exacerbate the challenges of infectious diseases. Politics are ubiquitous and may — much to the chagrin of subject matter experts — undermine a scientifically-driven

TABLE 2.1 Urban risk factors contributing to proliferation of infectious diseases.

Risk category	Risk factors
Environmental	Climate change
	Globalization
	Crowded living conditions
	Building infrastructure
	Land use
Social	Poverty
	Racial disparities
	Gender inequality
	Illiteracy
	Increased population mobility
	Underlying health conditions
	Vaccination
Political	Political will
	Advocacy
	Allocation of resources
	Governance

evidence base for decision-making resulting in a muck of propaganda, finger-pointing, and poor outcomes. Finally, injustices and discrimination abound when different groups perceive others as being the source of problems. Preparing for and responding to infectious disease events in cities, particularly New York City, offers challenges and opportunities unlike anywhere else.

Since 2018, NYC has reported at least five unique infectious disease outbreaks/clusters including measles, Legionnaire's disease, West Nile virus, and seasonal influenza [4]. With a strong public health and healthcare delivery landscape, however, these outbreaks were quickly addressed through a combination of surveillance activities for infectious diseases, screening and isolation, laboratory testing, outbreak investigation, intervention and control measures, ongoing education and training, and close collaboration with healthcare delivery partners.

The foundation for preparedness: education, training, and exercises

Education, training, and exercises are the foundation for preparedness. Conducting discussion- and operations-based drills and exercises of varying scope and intensity allow for facilities and healthcare delivery systems to test plans and processes for emerging and re-emerging infectious disease threats. With 4.4 million clinic visits and 1.1 million emergency department visits across NYC Health + Hospitals every year [5], it is critical for staff at all service lines — acute care, post-acute care, and ambulatory care — to understand the validity of infectious disease threats and the importance of preparedness, and to ensure that they are provided with the training, tools and resources required for rapidly and effectively responding to an infectious disease incident.

In-service training

Education is a most vital tool in infectious disease emergency management. A fundamental understanding of infectious diseases, risk assessments, basic infection prevention efforts, and specific institutional processes is an essential component of preparedness. Policies and procedures should build upon the knowledge and skills directly related to the job descriptions of various healthcare departments. By conducting routine in-service training, healthcare personnel are able to understand their role in the system and the importance of their actions during emergencies. Infectious disease emergency management in-service training should not replace existing education related to infection prevention or emergency management but should combine the two subjects to convey concise and direct information. Topics discussed in this type of training may include an overview of infectious diseases and special pathogens of high-priority or concern, patient screening strategies at points of entry, process flow from a point-of-entry to isolation, a review of transmission-based isolation precautions and infection prevention strategies, and other easily accessible tools and resources for healthcare personnel, such as internal policies or public health guidance.

With education serving as an essential component of preparedness, we recommend refresher courses and invite subject matter experts to speak on relevant topics, especially during outbreaks — whether they are localized outbreaks, epidemics, or pandemics — to ensure that healthcare personnel remain engaged and aware as guidance changes. For example, the System-wide Special Pathogens Program offers regular in-service training, both in-person and virtually, for the healthcare system. In addition, a monthly special pathogens meeting is held with a broad audience and open to all staff to provide updates on clinical guidance and processes and to discuss the latest infectious disease threats at the local, national, and international levels.

Mystery patient drills

Mystery patient drills, also known as secret shopper drills, allow healthcare facilities to assess preparedness and screening protocols. At NYC Health - + Hospitals, as well as at numerous institutions around the globe, healthcare staff are trained on three simple mitigation actions to proactively manage patients under investigation: identify, isolate, and inform. Operationalizing policies related to mitigation actions via mystery patient drills allows for strengths, gaps, and opportunities for improvement to be identified.

NYC Health + Hospitals System-wide Special Pathogens Program has conducted numerous mystery patient drills across the healthcare system in relation to Zika virus, measles, Ebola virus, and most recently COVID-19. Key data points revolving around screening, isolation, and notification to pertinent facility departments, and various infection prevention details and safety measures are collected throughout the drill, each with timestamps and detailed narratives documenting the experience from a patient actor. A variety of data points during these drills to better understand how existing processes are applied in a mock scenario (Table 2.2).

TABLE 2.2 Examples of data points collected during mystery patient drills at NYC Health + Hospitals.

Category	Data points
Simulation time stamps (time in minutes)	Arrival to registration area Arrival to donning of mask Registration to isolation Total duration (i.e., start to end of drill)
Entry and screening (yes/no)	Visible signage for travel history and symptoms throughout facility Masks conveniently located in waiting area Hand hygiene supply conveniently located in waiting area Patient asked about symptoms (fever, cough, rash) at registration Patient given a mask and instructed on how to wear it Patient asked to sanitize hands Patient asked about travel history Expedited registration process to limit patient contact
Isolation (yes/no)	Designated isolation room available Patient placed in private room with door closed Appropriate infection control signage placed on door Staff entering room wears appropriate personal protective equipment Personal protective equipment and supplies available near room entrance
Notification (yes/no)	Infection prevention notified Nursing or medical supervisor notified Facility leadership notified Department of health notified
Demonstrated staff capabilities (yes/no)	Ability to identify Ability to isolate Ability to inform

Mystery patient drills are generally followed by a debrief meeting involving those who designed the drill, departmental leaders, and individuals who tracked the data of interest. By debriefing, the facility leadership is able to understand and evaluate the effectiveness of existing strategies and processes, as well as the patient experience. This behind the scenes look into process flow from the angle of a patient allows for the facility to update or amend existing processes ranging from point-of-entry to triage and isolation, thereby leading to viable and sustainable policies. Mystery patient drills should be conducted on a routine basis and year-round to ensure staff are aware of facility policies and understand how to implement mitigation actions, especially during outbreaks or epidemics. Whether it is patient actors presenting with a painted rash simulating measles infection, or travel history combined with Ebola-like symptoms, these drills serve as a highly impactful tool in assessing existing safety measures, ensuring staff awareness, and reinforcing the importance of preparedness.

Full-scale exercises

To test the readiness of our health system, the System-wide Special Pathogens Program plans and executes full-scale exercises. These exercises, which can be stand-alone events or expand upon mystery patient drills, involve planning and designing a multi-agency exercise with a broader scope to test emergency operations plans related to infectious diseases. Various agencies such as local and state health departments, ground and aerial transport, police and fire departments, public health laboratories, and multiple healthcare facilities may partake in a full-scale exercise to test the ability to identify, isolate, inform, clinically manage, and transport one or more patients with a potentially highly infectious disease. Full-scale exercises may target a range of objectives such as the ability of healthcare providers to notify facility leadership and contact the local department of health to the overall response time from patient presentation to transport to a dedicated hospital. These multi-faceted drills test not only the ability of the health system to respond to an infectious disease threat, but also assess the infectious disease response infrastructure at the local- and regional-level.

It is crucial to consider realistic complexities that agencies or healthcare facilities may face when designing a full-scale exercise. For example, how would a hospital isolate and care for a parent and child who both are under investigation for a highly infectious disease such as Ebola? Adult and pediatric providers would need appropriate training on clinical care and infection prevention within a biocontainment unit, but must also understand the communication challenges that come with separating a scared child and a concerned parent, both of whom are ill. Additionally, if multiple patients present to an outpatient clinic, or even a long-term care facility, how would staff isolate symptomatic patients and visitors with a suspected airborne transmissible disease and a lack of airborne infection isolation rooms? These complex scenarios require detailed and strategic planning. Moreover, incorporating elements of surprise, or injects, such as fake vomit or a laboring patient, provide opportunities to practice managing unique events.

Another complexity is that of ensuring empathy to afraid, agitated, worried patients or those with communication challenges, such as language barriers or hearing and speech impairments. With unrecognizable healthcare workers wearing layers of protective gear, it is no surprise that patients may feel nervous or restless while isolated. Ensuring clear, consistent, and empathetic communication between clinical, operational, and administrative sectors across all agencies is critical in a high-stress environment. In addition, urban areas are often melting pots of diversity. Ensuring timely, relevant, culturally-sensitive risk communication is profoundly important. For example, during the 2015−16 Zika outbreak, communicating the risk of travel to pregnant mothers who were considering travel to areas of active Zika transmission such as Latin America and the Caribbean was important. Developing multi-language fact sheets and resource documents helped the community understand the risks. During mystery patient drills or full-scale exercises, for example, one of the data points collected are resources (e.g., disease-specific fact sheets) given to the actor/patient during clinical interaction.

Full-scale exercises generally begin with a pre-briefing meeting prior to the start of the exercise to align all participants with the targeted objectives, review the master scenario event list, and ensure all participants − whether physically present at the mock incident or within the emergency operations center − are aware of each simulated incident, injects, or other details related to the drill. Agency and facility specific after action reports and improvement plans must be developed after the exercise, using candid feedback, observations, and general performance outcomes related to specific objectives to address strengths, gaps, and opportunities for improvements.

To test infectious disease emergency operations, the System-wide Special Pathogens Program conducts, at least annually, city-wide full-scale inter-agency exercises on topical special pathogen threats. For example, in 2016, the full-scale exercise focused on Ebola virus disease. In 2017, the full-scale exercise was on Middle East Respiratory Syndrome, viral respiratory illness causing ongoing clusters of disease on the Saudi Arabian Peninsula. In 2019, in light of the Ebola outbreak in the Democratic Republic of Congo, NYC Health + Hospitals System-wide Special Pathogens Program led a multi-agency full-scale exercise. Four patient actors, both pediatric and adult, presented to various service line facilities within the healthcare system with Ebola virus disease-like symptoms and a relevant travel history. The scope of this exercise included two acute care hospitals, one post-acute care facility, one ambulatory care facility, along with the New York City Fire Department and New York City Department of Health and Mental Hygiene. Key objectives within this exercise revolved around identification and isolation strategies at unsuspecting locations, escalation and notification to public health authorities, travel time from a distant hospital to the Regional Ebola and Special Pathogens Treatment Centers, and communication to the emergency operations center. This tailored exercise highlighted the realism of patient presentation at any type of facility.

Overall, full-scale exercises allow healthcare facilities and agencies to improve coordination, collaboration, and communication, clarify responsibilities, and identity planning strengths, gaps, and opportunities for improvement during an infectious disease emergency, thereby leading to heightened preparedness and confidence in mitigation and response efforts.

Emergency management approach to infectious disease preparedness and response

Under the tenets of emergency management to reduce the vulnerability to all-hazards — whether natural, intentional, or technological in nature — NYC Health + Hospitals System-wide Special Pathogens Program focuses on preparing for and responding to emerging public health threats. The unique coupling of emergency management with special pathogens preparedness provides an array of tools and resources that can be utilized to respond to infectious disease incidents. The three tenants of communication, coordination, and collaboration, and their application, are critical for not only preparing for but also responding to urban health emergencies (Table 2.3).

Under the constant threat of emerging and re-emerging infectious diseases in a highly globalized society serving a very mobile patient population in NYC, the System-wide Special Pathogens Program's all-infectious disease approach has utilized the foundation of emergency management and applied the tools and resources to successfully respond to measles, Ebola virus disease,

TABLE 2.3 Basic tenets of emergency management.

Tenet	Application
Communication	Use of a mass notification platform to inform key stakeholders rapidly
Coordination	Use of an automated incident management platform to optimize information sharing, house protocols and processes, and provide a common operating picture across the enterprise
Collaboration	Access to a state-of-the-art emergency operations center to coordinate response efforts, collaborate with city, state, national and international entities, allocate and deploy resources, monitor and share information and gather key stakeholders to centralize and rapidly make decisions.

Zika virus, seasonal influenza, and *C. auris*. Disease-specific approaches are then customized depending on epidemiology, pathophysiology, and clinical requirements of the disease.

Response to Ebola virus disease

Ebola virus disease is a severe viral infectious disease affecting humans and nonhuman primates. From 2013 to 2014, NYC Health + Hospitals/Bellevue, an acute care facility and the Region 2 Ebola & Other Special Pathogens Treatment Center designated by the United States Health and Human Services in 2014 to care for patients suspected or confirmed to have Ebola or other special pathogen infections, received 20 persons under investigation for Ebola virus disease, including one confirmed patient. The emergency management structure applied for this special pathogen incident included a system-wide approach with hospitals, ambulatory care sites and even post-acute facilities receiving Ebola-specific training and resources (e.g., personal protective equipment, algorithms, plans, and processes). An Ebola concept of operations plan (ConOps) was developed and used continuously to maintain a state of readiness (Table 2.4). The NYC Ebola experience demonstrated the importance of urban healthcare delivery systems' ability to safely identify, isolate, transport, and treat patients with Ebola and other high-consequence pathogens. Using the tenets of emergency management, NYC Health + Hospitals conducted a full-scale interagency full-scale exercise on Ebola as well as a series of mystery patient drills across all its 11 acute care hospitals; continuously

TABLE 2.4 Ebola concept of operations plan (ConOps) content.

Key system elements	Emergency management
• Tiered healthcare facility strategy • Identification, isolation, & preliminary management at frontline hospitals, designated facility, & ambulatory/outpatient facilities • Ebola treatment facility transportation & intra-system transportation • Mobilizing special pathogen and preparedness teams • Laboratory services • Ebola-specific job hazard analysis • Mobilizing staff • Monitoring of healthcare workers • Public health monitoring and movement • Personal protective equipment resources • Environmental services and regulated medical waste management • Mortuary services	• Central office emergency operations center • Emergency operations center activation • Staffing guide • Resource allocation

communicated internally with all staff through system-wide town halls and externally with public health partners on the latest city-wide updates and coordinated planning and response efforts with city and state public health agencies. Through strong public health and healthcare delivery partnership, and through a coordinated public health emergency response, NYC safely and effectively tracked, monitored, and treated all suspected and the single confirmed patient with Ebola virus disease.

Using a similar ConOps concept, NYC Health + Hospitals published a Frontline Hospital Planning Guide for Special Pathogens [6]. This guide provides high level planning support and resources for identification, isolation, and preliminary management for patients with special pathogen diseases.

Response to the Zika virus

The response to Zika virus was the result of an increase in infections in the Americas, which resulted in widespread travel-associated cases in the United States. In 2016, there were 987 travel-associated cases of Zika in NYC [7]. The emergency management response to Zika virus in NYC proved to be one of the most challenging and complex efforts in recent years taking into account a swarm of clinical and epidemiological factors such as travel history, vector control, virology and testing capabilities, clinical management during pregnancy, congenital anomalies (e.g., microcephaly in infants), sexual and reproductive health concerns, and a susceptible population. These overlapping intricacies required a strategic response plan inclusive of experts from diverse and specialized fields.

Addressing the threat of Zika virus within healthcare facilities required tailored response and action plans. These tailored plans included a variety of considerations including universal travel screening protocols, symptom criteria and methods to differentiate Zika virus from similar vector-borne diseases, educational materials, signage in multiple languages depicting areas with active Zika virus transmission, clinical guidance and epidemiologic evaluation for Zika virus exposure, and diagnostic testing for Zika virus. Further, some screening criteria embedded referral mechanisms to ensure that high-priority patients, such as pregnant individuals, were promptly seen by specialists.

To align with the tenets of emergency management, response plans should be thoughtfully reviewed on a constant basis with facility leadership, subject matter experts, and public health partners to ensure that all processes reflect the most current recommendations. Maintaining situational awareness and transparency among the entire healthcare delivery system, or even throughout a single facility, will directly correspond to a prompt response. Consistent and scheduled briefings aid in maintaining situational awareness and should involve a multidisciplinary team to provide subject- and department-specific updates, all of which should be taken into account when revising or updating a response plan.

The NYC Health + Hospitals Zika Action Plan details the criteria for testing and reporting of Zika virus and is inclusive of screening algorithms for pregnant, non-pregnant females, males, pediatrics and neonatal populations based on local and national public health guidance (Fig. 2.1) [8].

Response to seasonal influenza

While the impact of seasonal influenza varies each year, it consistently places a significant burden on healthcare delivery systems. Annually, across the

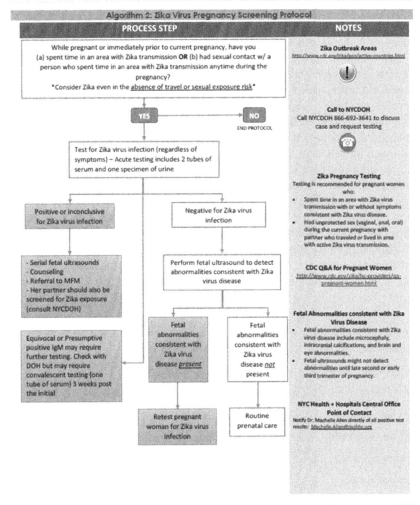

FIG. 2.1 Example algorithm for Zika virus pregnancy screening included in the NYC Health + Hospitals Zika Action Plan.

United States, hospitalizations associated with seasonal influenza can vary between 140,000 and 810,000 [9]. The NYC Health + Hospitals System-wide Special Pathogens Program uses an emergency management approach to effectively prepare for and monitor the impact of seasonal influenza. The approach sets a foundation for an efficient response by establishing mechanisms for activating a protracted, system-wide emergency operations center for seasonal influenza — tracking the impact of all influenza or influenza-like-illness (ILI) across 24 clinical care sites and providing additional support, as needed. As a part of this process, an incident management team is established at the system-level prior to the onset of flu season and includes specialists from a variety of fields and settings including virology, pharmacy, logistics and supply chain management, medical laboratories, infection control and prevention, and patient quality to support the protracted activation and provide guidance as the season unfolds.

Regardless of the size of a healthcare delivery system or organization, tracking qualitative and quantitative data on a consistent basis allows for heightened preparedness and a prompt response to any potential impacts of an outbreak on staffing, supplies, and space. Metrics that are tracked may range from status reports of ILI presentation to the emergency department, to the number of patients on droplet isolation precautions for suspected and confirmed influenza, to intensive care unit (ICU) impacts (Table 2.5). To track and monitor ongoing cases and potential impacts, incident management briefings occur as often as needed — such as weekly, biweekly, or even daily — depending on the severity of the flu season. An automated incident management platform may also be used to rapidly submit, analyze, and share summaries to clinical leaders for resource allocation and support.

Ultimately, providing routine updates on departmental impacts or concerns and leaning on subject matter experts to respond allows for healthcare facilities in NYC to maintain vigilance during flu season.

TABLE 2.5 Qualitative and quantitative metrics collected and tracked during flu season at NYC Health + Hospitals.

Qualitative metrics	Quantitative metrics
ILI emergency room impact	Total number of influenza tests ordered
ILI intensive care unit impact	Total number of lab-confirmed influenza cases
ILI facility impact	Total number of patients on droplet precautions
Flu supply status	Total number of healthcare acquired influenza cases
Staffing status	Employee vaccination rate
Impact to staffed beds	

Response to *Candida auris*

C. auris is an emerging multi-drug resistant fungus that has impacted numerous countries around the globe and several major cities within the United States. Within the United States, New York State has some of the highest numbers of *C. auris* cases — with 496 clinical cases as of March 2020 [10]. With high transmissibility, extreme drug resistance, and difficulties in prompt laboratory identification, *C. auris* has proved to be a threat that many healthcare organizations may find difficult to prepare for or respond to.

For example, delays in communication both within and between healthcare facilities, a lack of prompt screening measures, such as insufficient laboratory testing capacity, long turnaround times for results, and inadequate travel screening and patient risk assessments, can lead to the inadvertent spread of *C. auris* within a healthcare facility or the community. Due to these challenges and variability in *C. auris* detection and clinical management, it is critical to leverage an incident management platform. Doing so allows for health authorities to track key data such as community and healthcare-acquired cases, patients admitted with a history of *C. auris* colonization or infection, the number of patients on contact isolation precautions for *C. auris*; and to monitor facility impacts, which allows for the prompt involvement and response by key departmental specialists. For example, an uptick of healthcare-acquired *C. auris* cases is noted, infection prevention and environmental services departments are better enabled to rapidly deploy teams to assess the impacted units and implement strict environmental disinfection protocols. Infection prevention and nursing teams may retroactively conduct surveillance to investigate exposures, cohorting discrepancies, or incomplete screening and medical chart review at intake.

In NYC, the NYC Health + Hospitals System-wide Special Pathogens Program uses an automated emergency management platform to analyze *C. auris* case counts and impacts reported by clinical care sites. If concerning trends or upticks in cases are noticed, the System-wide Special Pathogens Program consults with key departments at the impacted facility to push forward an appropriate response, including but not limited to enhanced surveillance, increased communication between facilities, and increased environmental cleaning and disinfection. It should be noted that while many urban healthcare delivery systems and healthcare facilities may utilize infectious disease surveillance software, it is the emergency management approach used by the Special Pathogens Program that provides a strong foundation for a robust response. By incorporating multidisciplinary subject matter experts into an incident management team, healthcare facilities are able to plan thoroughly and respond swiftly no matter the infectious disease of concern.

Responding to the 2018–19 NYC measles outbreak

The 2018–19 NYC Measles Outbreak was the largest measles outbreak in the United States since 1992. In October 2018 an unvaccinated child from

Brooklyn acquired measles in Israel; this event, coupled with other importations of the disease from Israel, the United Kingdom, Ukraine, and other areas in the United States allowed for a small outbreak to rapidly grow in NYC. In less than 12 months, the outbreak resulted in over 21,000 exposed persons and 654 confirmed cases, 21 of which were acquired in healthcare facilities [11]. A majority of the exposures to measles occurred in medical facilities highlighting the challenges associated with rapidly identifying, isolating, and implementing infection control measures, as well as those relating to notifying public health authorities in a timely manner. Factors associated with these exposures are summarized in Table 2.6.

The measles outbreak required addressing the contagion on two fronts: first, the contagion itself. Including rapidly identifying and isolating patients with suspected measles, testing, counseling and caring for patients. And second, responding to the contagion of misinformation which ran rampant and deep in NYC. This included vaccination hesitancy, and misinformation, anti-vaccine sentiments, and general disease misinformation.

Responding to the contagion: the disease

The response in NYC included the pre-screening of patients presenting to healthcare delivery sites (e.g., emergency department, clinics) for symptoms and immunity status through the use of rooms and entrances separate from general waiting rooms, as well through telephone pre-screenings before scheduled visits. Within the healthcare setting, triage activities required prompt screenings which included a brief questionnaire used for rapid triage and isolation based on symptomatology (e.g., fever and rash, fever and possible measles exposures, fever and travel from a geographic area with a measles outbreak, fever and cough, standalone cough, coryza, or conjunctivitis, and prodromal of fever). The use of electronic medical health records (EMR) to automate the screening, exposure history, and travel history reduced triage time and allowed for faster patient isolation (Fig. 2.2). In addition, built-in best practice advisory prompts helped clinicians implement infection control and prevention interventions including placing patients in airborne infection isolation rooms (AIIR), source control (e.g., providing a mask to the patient), and automated order sets to facilitate process and escalate clinically as warranted. This leveraging EMRs was crucial in the identification, isolation and notification of infectious diseases patients.

Additionally, staff education played a vital role in addressing the measles outbreak. Public health guidance continued to change according to the latest epidemiologic information, new public health regulations, vaccination requirements, clinical guidance, and infection control and prevention interventions. Ongoing education on the hierarchy of controls, outbreak specific clinical guidance and evolving geographical localities of disease transmission assisted with the local measles response. Even clinical guidance that wasn't

TABLE 2.6 Unique challenges regarding the measles response and innovative strategies used by NYC Health + Hospitals.

Consideration	Challenges	Innovative strategies
Engineering controls	- Limited airborne infection isolation rooms (AIIR)	- Use of portable HEPA filters - Prioritize existing AIIR for suspected/confirmed measles patients - Converting existing units to negative pressure rooms
Clinical presentation	- Exposures before rash onset - Often non-specific initial signs and symptoms	- Prominent signage and posters in multiple languages on key measles signs and symptoms strategically placed for patients to self-report to front desk - Respiratory stations with facemask, tissues, hand sanitizers for patient use for source control - Ongoing education to clinicians for high index of suspicion and to recognize measles in the prodromal phase - Prompt triage screening through use of symptom-based and travel screening for clinical escalation
Administrative controls	- Risk communication - Measles immunization status in healthcare workers - Tracking of suspected and confirmed measles cases - Preventing nosocomial spread of measles	- Occupational health providing departments with lists of non-immune staff, and real-time notification to infection control of all suspect cases - Ongoing education sessions and huddles for clinicians on latest outbreak news and clinical guidance - Use of an automated data collection platform to track suspected and confirmed measles patients - Restricted or limited visitors to high risk units (ICU, Labor & Delivery, Neonatal Intensive Care Unit) unless proof of immunity
Outbreak-specific controls	- Vaccine hesitancy and anti-vaccination propaganda - Misinformation on measles outbreak	- Ongoing education to providers and providing the tools and resources to combat misinformation - Patient pamphlets and resources to help educate the community on disease facts

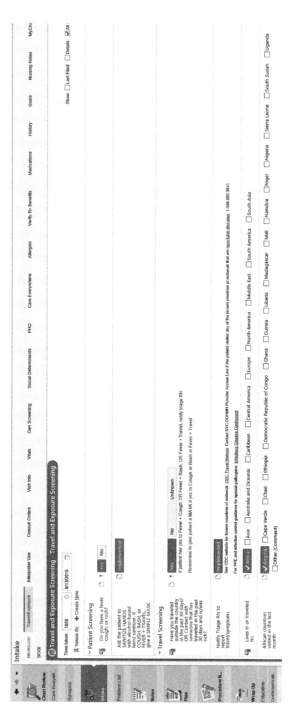

FIG. 2.2 Electronic medical health records travel and exposure screening.

new (e.g., infected persons can spread the disease from four days before rash onset through four days after), ensuring clinicians have a high degree of suspicion with prompt isolation is key, and required ongoing education.

Responding to the contagion: misinformation

The spread of misinformation plays a critical role in outbreak response. The social communication and behaviors of the public can make or break public health interventions, such as promoting vaccinations. Therefore, ensuring the public has the most up-to-date and accurate information is paramount. The NYC Measles outbreak was propagated not only multiple importations of the disease from other countries, but also prevalent delays in vaccination and anti-vaccination sentiments. For example, over 70% of the measles diagnoses in adults aged 18 years and older were unvaccinated or had unknown vaccination history [12].

Misinformation seen in the NYC measles outbreak included handbooks, pamphlets, fliers, hotlines, and anti-vax gatherings that contradicted the medical and scientific consensus that measles is a real threat and vaccines are safe and effective − all of which contributed to vaccine hesitancy, that can manifest in delaying vaccination or not vaccinating at all.

Healthcare delivery sites, such as hospitals and clinics, played a crucial role in providing health education and keeping communities informed. As patients presented for clinical care service and support, providers were tasked with answering questions and alleviating concerns including addressing misinformation and vaccine hesitancy. To ensure that healthcare providers received up-to-date information and ongoing education as it related to the measles epidemic, various educational services were offered through a multi-pronged approach. This included daily updates in form of "departmental huddles," webinars, town hall events, emergency operations center activation conference calls, incident briefings, all staff emails and clinical council correspondences. These modes of communication were supplemented with ongoing efforts by public health partners to further engage providers in impacted communities through ongoing health alerts, frequent webinars, and continuous provider calls.

Besides educated clinicians on the frontlines of battling the measles epidemic, it was vital to provide educational resources to the patient population and overall community. NYC Health + Hospitals developed and shared numerous educational pamphlets, posters, and signages in multiple languages on all aspects of measles including basic disease facts, advice if you have measles, staying safe from getting measles, keeping family and others around you safe, the safety and efficacy of vaccinations, and other pertinent topics (Fig. 2.3).

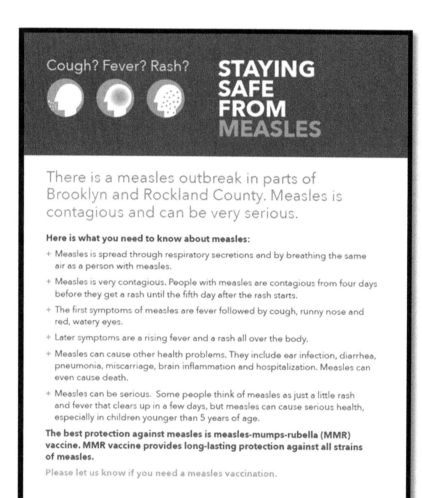

FIG. 2.3 NYC Health + Hospitals measles prevention patient pamphlet used during the 2018–19 measles outbreak.

Conclusion

New Yorkers, White says, do not crave comfort and convenience, for if we did, we'd live elsewhere [12], while Steinbeck insists that once you have lived in New York City, and it becomes your home, no place else is good enough [13].

Wherever one may lie on this spectrum, the history, energy and prospects of its urbanity are undeniable and so too, are its challenges. Ciselak, Herstein, and Kortepeter note that recent decades have brought communicable disease outbreaks caused by novel or emerging pathogens with the potential to bring significant morbidity and mortality in humans [14].

Taken together, the approaches used by NYC Health + Hospitals' System-wide Special Pathogens Program allow for it to continually adapt to the ever-changing infectious disease landscape with innovative training and response resources to ensure that patients receive the highest quality and safest care available in a global city. From measles to Ebola to Zika, infectious disease threats are not rare or isolated events. In fact, in our highly globalized world we can and should expect more infectious disease outbreaks spreading faster and farther than ever before. As such, all healthcare systems, but especially those in urban environments, must continue to invest in special pathogens preparedness, continue educational and training efforts for healthcare personnel, educate the public on the importance of immunizations for individuals and communities, and work to maintain a state of readiness. After all, "ready or not, patients will present."

References

[1] New York City department of health and mental hygiene. Population - New York City population. 2019. www1.nyc.gov/site/planning/planning-level/nyc-population/population-facts.page. [Accessed 27 October 2020].

[2] Jacobs J. The death and life of great American cities. New York: Random House; 1961.

[3] Beauregard R. Cities in the urban age: a dissent. Chicago: University of Chicago Press; 2018.

[4] New York City department of health and mental hygiene. Health Alert Network (HAN); 2019. www1.nyc.gov/site/doh/providers/resources/health-alert-network.page. [Accessed 27 October 2020].

[5] New York City Health + Hospitals. About NYC Health + Hospitals. 2019. www.nychealthandhospitals.org/about-nyc-health-hospitals/. [Accessed 27 October 2020].

[6] NYC Health + Hospitals emergency management, NYC Health + Hospitals special pathogens Program. Frontline hospital planning guide: special pathogens. New York: NYC Health + Hospitals; 2019.

[7] New York City department of health and mental hygiene. Zika virus. 2019. www1.nyc.gov/site/doh/health/health-topics/zika-virus.page. [Accessed 27 October 2020].

[8] Madad SS, Masci J, Cagliuso Sr NV, Allen M. Preparedness for Zika virus disease − New York City. MMWR 2016;65:1161−5.

[9] Centers for Disease Control and Prevention. Disease burden of influenza. 2019. www.cdc.gov/flu/about/burden/index.html. [Accessed 27 October 2020].

[10] New York State Department of Health. Get the facts about *Candida auris*. 2019. www.health.ny.gov/diseases/communicable/c_auris/. [Accessed 27 October 2020].

[11] New York City Department of Health and Mental Hygiene. Alert #10: update on measles outbreak in New York City and citywide recommendations. 2019. www1.nyc.gov/assets/doh/downloads/pdf/han/alert/2019/update-measles-outbreak.pdf. [Accessed 27 October 2020].

[12] White EB. Here is New York. New York: Harper & Bros; 1949.

[13] Steinbeck J. America and Americans. Portsmouth, New Hampshire: Heinemann; 1966.

[14] Cieslak TJ, Herstein JJ, Kortepeter MG. Communicable disease and emerging pathogens: the past, present and future of high-level containment care. In: Hewlett A, Murthy ARK, editors. Bioemergency planning: a guide for healthcare facilities. Cham: Springer International Publishing; 2018.

Chapter 3

The 2019 measles outbreak in Clark County, Washington

Michael A. Stoto[1], Rachael Piltch-Loeb[2], Roxanne Wolfe[3], Robin Albrandt[3], Alan Melnick[3]

[1]*Department of Health Systems Administration, Georgetown University, Washington, DC, United States;* [2]*Emergency Preparedness Research Evaluation & Practice Program, Harvard TH Chan School of Public Health, Boston, MA, United States;* [3]*Clark County Public Health, Vancouver, WA, United States*

In early 2019, Clark County, Washington and neighboring counties in Washington and Oregon experienced a measles outbreak involving a total of 78 confirmed cases that challenged local and state public health authorities in a number of ways. Clark County Public Health (CCPH) led the response in collaboration with public and private entities in Clark County, Washington, and Oregon. Over the course of four months, CCPH coordinated activities to control the outbreak including active surveillance and monitoring, implementation of social distancing and quarantine policies, and emergency risk communication via traditional and social media. Although successful in many ways, the response of CCPH and partners to the outbreak illustrates some important challenges in public health emergencies and successful approaches to addressing them. One challenge was the tension between publicly identifying the community in which the early reported cases occurred to aid in the outbreak investigation and help people understand what was going on and at the same time avoiding stigmatizing the community and perpetuating a misunderstanding of risk that others faced. Another was responding to false claims on social media, many from outside Clark County, in a way that allowed CCPH to identify real community concerns and demonstrated that the health department was willing to engage in a dialog with community members who had these concerns.

The public health response required assembling an incident management team including hundreds of individuals that identified public exposure sites, traced hundreds of susceptible contacts of measles cases, quarantined them, and monitored them daily. Clark County Public Health also worked with public and private schools and child care centers to exclude hundreds of

Inoculating Cities. https://doi.org/10.1016/B978-0-12-820204-3.00003-6
41

children at risk of infection. Managing this effort required effective coordination and communication among many individuals and entities: the CCPH and the Washington State Department of Health (DOH) incident management teams and others in CCPH, the DOH, other state and local public agencies, the schools, and the private healthcare system. The response also required the coordination of physicians, nurses, epidemiologists, emergency personnel, educators, and other disciplines.

This chapter begins with a background section describing the health threat of measles and the public health emergency preparedness system in Clark County and Washington State. The methods section that follows describes the "facilitate look-back" approach used to prepare an after-action review, on which this chapter is based. The analysis section covers policy decision-making, emergency risk communication, outbreak management, and coordination and communication. The conclusions draw on Clark County's experience to identify lessons about both public health strategies and emergency management that are potentially applicable to public health emergencies caused by infectious disease outbreaks in other settings.

Background

Measles

Measles is a highly contagious virus, transmitted through the air when an infected person coughs or sneezes. The virus is characterized by a rash, and can cause fever, runny nose, and watery eyes. Measles can lead to significant complications especially for high-risk groups including children, pregnant women, and individuals with compromised immune systems. As many as one in twenty children with measles is at-risk for pneumonia, the most common cause of death from measles in young children; while rarer, but deadly, complications also include encephalopathy (brain damage or malfunction).

Since 1978, there has been a robust vaccination campaign in the United States to mitigate the spread of measles, leading to the disease successfully being eliminated in 2000. The United States held elimination status for measles for nineteen years as a result of a successful vaccine and significant public health efforts [1]. However, in 2019, the elimination status was threatened. According to the United States' Centers for Disease Control and Prevention (CDC), 2019 saw the greatest number of cases of measles in the United States since 1992. A total of 1,282 individual cases were confirmed across 31 states; including 128 who were hospitalized. The majority of cases in 2019 were among those who were not vaccinated for measles [2].

Typically, the public health strategy for controlling the spread of the measles virus relies upon near-uniformity in vaccinating children. Herd immunity is a population-level threshold that limits the likelihood of epidemic transmission of a viral agent and epidemiologists estimate that herd immunity for measles requires comprehensive vaccination of at least 95% of a suscep-tible population [3]. However, a rise in vaccine hesitancy — the reluctance or refusal to vaccinate — has led to a corresponding reduction in vaccination coverage. Indeed, globally, vaccine hesitancy has become one of the greatest threats to public health, according to the World Health Organization which named it one of the top ten threats to health in 2019 [4].

In the United States, immunization rates which are reported through state immunization records suggest at the state level, children under five have maintained near 95% vaccine coverage. In Clark County, however, only 78% of children aged 6—18 years have had two doses of the measles, mumps, and rubella (MMR) vaccine in 2018 [5].

Population data, however, obscures the variation in vaccination coverage among particular sub-groups. Whereas the National Immunization Survey reported that 94.3% of US children entering kindergarten in 2017 had at least two doses of the measles vaccine, recent measles outbreaks have been pro-pelled by population clusters whose vaccination rates were considerably lower. In a measles outbreak among Somali-American children in Minneapolis in 2017 the vaccination rates among that immigrant community had plummeted from 94% in 2004 to 36% in 2014 [6]. Other highly susceptible population clusters have been identified in the recent outbreaks, including an Ohio Amish community as well as pockets of progressive parents who were skeptical of the vaccine's safety [7].

During January 1—October 1, 2019, a total of 1,249 measles cases and 22 measles outbreaks were reported in the United States. Even though only the first nine months of the year, this represents the most US cases reported in a single year since 1999, and the second highest number of reported outbreaks annually since measles was declared eliminated in the United States in 2000. Outbreaks in Orthodox Jewish communities in New York state (i.e., Rockland County and New York City) accounted for 934 (75%) cases during 2019, and 89% of all cases were in those who were unvaccinated [8]. Robust responses in these locations with multiple partners involved the administration of approx-imately 60,000 MMR vaccine doses in the affected communities; tailored communication campaigns; partnerships with religious leaders, local physi-cians, health centers, and advocacy groups; and use of local public health statutory authorities. Nevertheless, these nearly year-long outbreaks threatened the elimination status of measles in the United States [8]. In 2019, vaccination rates in Rockland County were estimated to be approximately 77% for all schoolchildren in the county, suggesting that the rates among children there are lower than the rates reported by the National Immunization Survey [9].

As a result of varied belief systems and state policies, vaccine hesitancy has a greater impact in particular communities. Depending on the state, vaccine hesitancy among parents is countenanced by a variety of waivers that permit children to avoid legally-required vaccinations and still enter school as kinder-gartners. These waivers typically fall into one of three broad categories: medical exemptions, religious belief waivers, and philosophical or personal-belief waivers. All states permit waivers for children with medical conditions that render them susceptible to vaccination harms. Thirty states permit religious-belief exemptions, and an additional eighteen states allow personal-belief exemptions as well [10]. Three states — California, West Virginia, and Mississippi — did not allow for either religious or personal-belief exemptions. During the measles outbreak in 2019, several states including New York, Maine, California, and Washington, passed new laws eliminating various exemptions for vaccines. In New York the state legislature eliminated religious exemptions, Maine removed both religious and philosophical objections, and California removed medical exemptions without the approval of a public health official [11–14]. Washington removed personal exemptions for the MMR vaccine after the outbreak in 2019; previously 7.5% of schoolchildren had exemptions: 5.9% personal, 0.9% medical, and 0.6% religious [15,16].

The New York City outbreak in particular was seen as a consequence of under-vaccination. Between September 30, 2018, and July 15, 2019, a total of 649 cases of measles were confirmed in the city. Initiated by repeated importation of cases from Israel, nearly all (93.4%) of the cases were members of the Orthodox Jewish community, and 91.5% resided in the Williamsburg or Borough Park areas of Brooklyn. Consequently, the Department of Health and Mental Hygiene's primary control measures focused on improving immunization. More than 20,000 contacts were identified, informed of their potential exposure, and referred for post-exposure prophylaxis with MMR vaccine, immune globulin, or home quarantine, as indicated. Unvaccinated children were identified with the use of the Citywide Immunization Registry and their parents contacted about making an appointment for MMR vaccination. The recommendations for MMR vaccination were revised for persons residing in, or regularly spending time in, neighborhoods in which ongoing measles transmission was present. On December 6, 2018, schools and childcare programs in Williamsburg and Borough Park were notified that all children without age-appropriate MMR vaccination or proof of measles immunity, including those with medical or religious exemptions, were to be prohibited from attending school and childcare programs. Three schools and nine childcare programs were closed temporarily under an order of the Commissioner of the Department of Health and Mental Hygiene. On April 9, 2019, an emergency order was issued requiring MMR vaccination or proof of measles immunity for all persons living, working, or going to school in the four affected Williamsburg ZIP Code, and 232 summonses were issued for not adhering to the emergency order. Already high antivaccination sentiments were deepened when an organization targeted this community with misleading materials regarding the risk of vaccination. To address these concerns, the Department of Health and Mental Hygiene reprinted and mailed two booklets that provided accurate information about vaccines to 29,000 households in Borough Park and Williamsburg. In addition, the public was notified through press releases, numerous media interviews, print ads in newspapers serving the Orthodox Jewish Community, social media, and robocalls to the affected neighborhoods in English, Yiddish, and Spanish [17].

The Washington State Public Health Emergency Preparedness System

Clark County is a county in the southwestern part of the US state of Washington, in the greater Portland, Oregon metropolitan area, and the southernmost county in Washington (Fig. 3.1). As of July 2018, the population was 481,857 [19], making it Washington's fifth-most populous county. The Clark County seat and largest city is Vancouver, which is located across the Columbia River from Portland. Approximately 24% of the population is under 18. More than 82,000 students are served by 150 public schools in 11 school districts and multiple private schools [20].

In the United States, public health primarily operates under state laws and regulations (as opposed to federal), but is implemented largely at the local level. Washington is a "home rule" state, so CCPH rather than the DOH had the lead in responding to the outbreak. Clark County Public Health has a staff of around 110 that provides services to "influence conditions that promote health, such as access to healthy and affordable foods, clean water, health care, and neighborhoods that are safe for walking and biking [and to] minimize the impact of disease outbreaks through vaccination efforts, early detection, and swift responses" [21]. Clark County Public Health also coordinates Washington state's Public Health Emergency Preparedness and Response Region IV in Southwest Washington, which includes four counties (Clark, Cowlitz, Skamania, and Wahkiakum) and the Cowlitz Tribe. Public health systems, however, are multi-jurisdictional and multidisciplinary. In addition to state and county health agencies, public health preparedness "systems" include partner

FIG. 3.1 Clark County, Washington [18].

agencies such as hospitals and physicians (most of which are not public agencies or employees), emergency medical services agencies, schools, and others who may not think of themselves as having a public health role [22].

Methods

Clark County Public Health and the Washington State DOH have conducted their own after-action reviews (AARs) of this local public health emergency. An epidemiological analysis has also been published [23]. These reports contain many useful lessons learned, especially about the operations of CCPH's and DOH's incident management teams. Experience in other public health emergencies, however, suggests that additional and deeper lessons about emergency preparedness systems can be drawn from a broader, systems-oriented review that involves both multiple public health agencies and other organizations and individuals that were involved in the response. Thus, with the intention of identifying lessons from the 2019 Clark County measles outbreak, the Emergency Preparedness Research, Evaluation & Practice (EPREP) Program at the Harvard T.H. Chan School of Public Health facilitated a joint AAR with participation from Clark County public health and emergency management, representatives of Clark County schools, and Washington State Department of Health.[1]

In order to conduct this review, the EPREP first held initial planning meetings with key personnel from CCPH. These meetings were followed by bi-weekly joint meetings with CCPH and the DOH to discuss key issues and the planning process. After this, two independent EPREP members reviewed the existing AARs prepared by CCPH and the DOH and discussed key issues that arose as they related to PHEP practices. Staff also reviewed press reports and other documents as a means of supplementing the AAR content. An initial list of challenges and decision-points was then compiled based on the document review. This initial list was based on the issues identified by EPREP reviewers and informed by input from CCPH and DOH personnel. Once the initial list of challenges was assembled, eight semi-structured telephone interviews with other key informants — including public information officers, incident commanders, epidemiologists, planning chiefs, operations chiefs, and those outside of the Incident Management Team (IMT) such as school leaders and healthcare providers —to identify based the roles that these individuals played in the outbreak response. At the conclusion of the telephone interviews, two EPREP members reviewed the interview notes and expanded upon the list of key issues and decision-points. No new issues were identified in this process, but previously identified topics were fleshed out in more detail and

---

[1]A list of AAR participants — including those from Clark County Public Health, the Clark County School System, and the Washington State Department of Health — can be found in Appendix 1 of the Notes section.

BOX 3.1 Steps conducted to complete the joint AAR in Clark County, Washington

(1) Hold planning meetings with key personnel;
(2) Review draft AARs prepared by CCPH and the DOH;
(3) Review press reports and other documents to supplement AAR content;
(4) Compile an initial list of challenges and decision-points based on the document review;
(5) Conduct telephone interviews with other key informants;
(6) Review interview notes to identify the most critical cross-jurisdictional issues and expand upon the list of key issues and decision-points;
(7) Facilitate discussion with key personnel to address issues and identify joint lessons learned.

---

prioritized. The process concluded with a facilitated discussion with CCPH and DOH personnel to address the key issues identified and articulate joint lessons learned (Box 3.1).

The discussion took place in September of 2019 in Vancouver, Washington. Harvard EPREP staff facilitated the discussion using the Facilitated Look-Back method developed by the RAND Corporation [24]. This approach brings together key stakeholders and responders in a meeting to probe dimensions of decisions and explore nuances in past decision-making. Using a neutral facilitator and a no-fault approach, the discussion focused on decision-making and the shared experience of the event rather than the responses of individual actors to solicit improvement strategies. The participants agreed to follow the Chatham House Rule (i.e., that nothing would be attributed to specific individuals).

There were sixteen participants in the facilitated lookback including five CCPH representatives, seven DOH representatives, one representative from Clark County emergency management, and three representatives from two of Clark County's two largest school districts. Their names and affiliations are listed at the end of the chapter.

The meeting lasted five hours and was conducted in two sessions. The morning session focused on policy decision-making and emergency risk communication. The afternoon session focused on outbreak management and coordination and communication. For each topic, several key questions were used to guide the group: what were the critical events & decisions, what went well and why, what could have gone better, what could have been done differently, what were the underlying problems, and what system changes are needed to improve future efforts. The results below are organized by these four topics and guiding questions.

Incident description

On December 31, 2018, a healthcare provider notified CCPH of a suspect measles case in an unvaccinated child who had recently arrived from Ukraine. On January 3, 2019, lab results confirmed that this child had measles. Eleven days later, on January 14, 2019, CCPH had reports of three confirmed cases of measles in children and 11 suspect cases. The next day, on January 15, 2019, CCPH delegated authority to the CCPH IMT to manage the measles outbreak. On January 18, 2019, Clark County declared a public health emergency to address the needs of this rapidly expanding outbreak. The DOH IMT was activated on January 15, and the Governor declared a state public health emergency on January 25, 2019.[2]

Over the course of the response, the CCPH IMT coordinated various activities to control the outbreak including active surveillance and monitoring, implementation of social distancing and quarantine policies, emergency risk communication with the public via traditional media and social media, and coordination with county schools and hospitals in Clark County. The DOH IMT supported CCPH with parallel activities at the state level and coordinated with state agencies, such as the Office of Immunization and Child Profile (OCIP), which manages the Vaccine for Children Program. Both IMTs communicated with the Oregon Health Division and other local jurisdictions both in Washington and Oregon to coordinate surveillance and other activities.

At the end of the response, CCPH had identified 71 confirmed cases, a majority of which were in unvaccinated persons under 18 years of age. There were 4,138 contacts who were traced and personnel made daily monitoring calls to 816 individuals. The IMT identified 53 public exposure sites that included 13 healthcare facilities, 15 schools and child care centers, one workplace, and 24 other public places. CCPH also worked with schools and child care centers to temporarily exclude 849 susceptible (un- or under-vaccinated) students in 13 schools in three public school districts and two private schools.

Over the course of the response, 237 individuals from CCPH, Washington State DOH, Medical Reserve Corp, CDC, interpreters, other health departments, emergency management, and other volunteers were involved. At any one time, between 30 and 40 CCPH staff participated in the response, and 89 of the 110 were involved at some point during the outbreak. Many others, especially at the state level and in Clark County schools, also participated in the response. The CCPH IMT was deactivated on March 18, 2019, after being in response for 63 days. The total cost to CCPH was over $864,000.

---

[2]A timeline of cases and Incident Management Team activities can be found in Appendix 2 of the Notes section.

Analysis

Policy decision-making

CCPH's strategy for responding to the outbreak differed from New York City's approach in two ways. First, in addition to tracing the contacts of confirmed cases to identify people who might have been exposed, CCPH identified and publicized sites where exposure might have taken place. This was done through press releases as well as directly for individuals who had been at a site where personal information was routinely collected, such as clinics and emergency rooms. Individuals who had been to these sites were informed about preventive interventions such as immunization or immune globulin, if applicable, and the need for quarantine for the duration of the incubation period. Public exposures such as in churches, schools, clinics, or child care centers were most common during the first four weeks of the outbreak and decreased following the community-wide implementation of CCPH-recommended outbreak control measures (e.g., social distancing and quarantine). Among the 30 patients identified after February 1, 26 (87%) were known contacts in quarantine and under active surveillance [23].

Second, because the outbreak was concentrated in a religious community that had a history of mistrust and vaccine hesitancy, New York City focused on vaccinating all children in two neighborhoods. Clark County, on the other hand, adopted a social distancing strategy for the entire county, excluding children and adults who were susceptible due to their vaccine status from schools with potentially-active measles cases and voluntary home quarantine of exposed individuals. As discussed below, there was a substantial increase in vaccination in Clark County during the outbreak, with most provided by children's own pediatric providers.

CCPH's decisions were made quickly and communicated effectively to schools for implementation at the school level. Moreover, this approach was successful in ending the outbreak in about three months. This compares favorably with contemporaneous outbreaks in New York City and State, which lasted about ten months [17].

One of the challenges that CCPH faced was whether to publicly identify the Slavic community in Clark County as a focus of the outbreak. The first reported case was a member of this community, and there was an early cluster of cases in this group. From a strictly epidemiological perspective, and in the view of some of CCPH's partners, publicly identifying this pattern might have helped to facilitate case finding and outbreak analysis. In addition, the national attention to the measles issues created pressure to associate the outbreak with a particular group. On the other hand, although the first diagnosed case was a child from this community who had traveled from Ukraine, epidemiologic data did not confirm this was the first case in the county. Furthermore, in contrast to the measles outbreak occurring in New York, where cases were mainly confined to the Orthodox Jewish community [17], Clark cases were not limited to the Slavic community. Rather, because vaccination rates were low throughout Clark County, everyone was at-risk. Thus, CCPH made a conscientious decision to avoid publicly identifying this community. This helped to avoid stigmatizing and appearing to blame the Slavic community, which might have inhibited their cooperation, as well as to not misrepresent the risk to the larger community. Maintaining the trust and cooperation of this community helped facilitate the rapid and effective identification of exposure sites and active surveillance, as well as with voluntary home quarantine.

Another success was the rapid initiation of emergency operations. CCPH and DOH stood up their IMTs on January 15, as soon as it was apparent that local transmission was occurring. On the same day, CCPH sent out its first provider health alert and media release, updated its website and social media, and initiated case notification and contact tracing. At this point, the Medical Reserve Corps was activated along with the Washington State local health jurisdictions' mutual aid agreement to help fill response staffing needs. Clark County declared a public health emergency three days later, and the state followed suit on January 25. These declarations allowed for increased staffing for CCPH, support for the county from the DOH and resources outside of Washington Public Health Emergency Preparedness and Response Region IV, as well as for expenses incurred in the response to be reimbursed. The state declaration also helped expedite the receipt of additional resources from outside the state through the Emergency Management Assistance Compact. As a result, CCPH could operate at a more robust level early on in the response. The emergency declarations also helped reduce resistance to social distancing procedures and increased support for the health officer from the Clark County Council.

Another challenge was that the measles outbreak brought prominence to legislation at the state level to remove philosophical exemptions for vaccines for children entering public school. This complicated the public health response, as DOH staff were asked to supply data in support or opposition to the bill. At the state level, the need to provide data to policymakers and for legislation contributed to a division in staff time that otherwise may have been entirely focused on the outbreak response. Meeting participants suggested that in future responses, specific individuals should be identified to respond to political inquiries and needs. These individuals should not be part of the IMT, but rather an administrative or subject matter expert who sits outside of the incident command structure. Delineating these roles would help alleviate the burden on the IMT, and, when personnel ask for information from other responders, their purpose and motivation for needing said data points will be clear.

Emergency risk communication

Media inquiries were handled by the CCPH IMT Public Information Officer, who worked with the Clark County health officer, who acted as spokesperson, to provide a clear consistent message. One challenge was to provide a narrative of the outbreak that did not identify or stigmatize the Slavic community for the reasons discussed above. For example, without any supporting evidence, some anti-vaccine groups spread the false narrative that this was an "inbreak" (i.e., contained in only one community) that the rest of the community didn't have to be worried about.

A vocal national anti-vaccination movement, along with state legislation to eliminate philosophical exemptions introduced during the outbreak, created challenges for emergency risk communication. There were many false claims on social media, many from outside Clark County. CCPH monitored social media and allowed vaccine-hesitant persons to post on CCPH's own Facebook page. Comments that demonstrated anti-vaccine sentiment were not deleted but rather were responded to with factual information. This allowed CCPH to identify some real community concerns or questions related to vaccines and demonstrated that CCPH was willing to engage in a dialog with community members who had these types of concerns rather than simply shut down the messages.

One of the key challenges for emergency risk communication was notifying schools and families about school exclusions. Both the schools and CCPH pointed to the effective communication between CCPH to the school systems to indicate which schools would need to implement the exclusion of children, teachers, and staff who were not up to date on their vaccines. Communication from CCPH to schools occurred through letters and contact with the school systems' administrators and public information officers.

Analysis of the measles outbreak suggests that additional work is needed with schools, the medical community, and community groups to build trust for future responses and to address vaccine hesitancy in general. In the context of community groups, there is a limited budget to engage in this work, so CCPH and the DOH should think creatively about ways to build trust through existing programs, before engaging in the vaccine hesitancy conversation. This could be through focusing on issues that are important to the community (e.g., chronic disease issues) to demonstrate the value of public health. Additionally, since many parents in this area do not have experience with measles, having conversations about more familiar conditions might be helpful. School nurses and existing community liaisons who speak the language of vaccine-hesitant parents also can be key resources.

Outbreak management

Active surveillance required CCPH to identify potential exposure sites. By the end of the outbreak 53 sites were identified, including 13 healthcare facilities, 15 schools and child care centers, one workplace, and 24 other public places. Once identified, the IMT worked with healthcare facilities to identify who was present 30 min before to two hours after the patient had left. Keeping track of confirmed cases, potential cases, exposed individuals, and exposure sites (some of which were in other jurisdictions) is a labor-intensive process that the Washington Disease Reporting System was not designed to manage. CCPH developed an *ad hoc* database within the first month of the outbreak to manage this process. It included a data dashboard that provided a daily snapshot of the outbreak to the IMT.

The school exclusion policy was successful but resource-intensive. A total of 849 unvaccinated students in 15 schools in three public school districts and two private schools were temporarily excluded. This created a burden on schools and the healthcare system, as well as the CCPH IMT, and managing this effort was challenging in several ways. Implementation required knowing the immunization status of all children, but records were not always up to date. County-level immunization records based on parental reports at school enrollment were not consistent with state-level immunization data based on a registry in which healthcare providers participate. Oregon, where some Clark children receive their healthcare, also has a vaccine registry, but the two state systems were not compatible. Adults such as teachers and bus drivers also had to be excluded if not immune. In order to be at work, these personnel had to demonstrate they had received the vaccine or had immunity to the disease based on their titers. Obtaining vaccination records for adults, especially from outside the state of Washington, proved challenging. For some adults who needed the vaccine, costs and availability were barriers.

Consistent with existing practice, CCPH encouraged that children be vaccinated by their own pediatric providers. This recommendation leveraged existing relationships and the capabilities of a large number of local and regional providers who already had the measles vaccine in stock. As a result, the number of immunizations in January and February increased nearly three-fold for children and 15-fold for adults. The increase was especially large in the two weeks between January 20 and February 2, that is, following the initial alerts of the outbreak. Providers had to be advised, however, that the measles vaccine should not be given more than 72 hours after exposure because CCPH would not know if a rash was the result of a benign vaccine reaction or an actual case of measles. Some providers familiar with the state OCIP through the Vaccine for Children Program and parents reached out to the OCIP during the outbreak about vaccination and other issues. The OCIP helped to facilitate access to the MMR vaccine, including for adults, for whom it is not typically administered.

Although this approach generally received support and approval, some individuals thought that public vaccination clinics should have been set up. Public clinics were eventually organized in conjunction with the Legacy Salmon Creek Medical Center, but not until March, when the outbreak was waning. The delay was due in part to limited coordination between CCPH, the DOH, and the medical center regarding staffing, standing orders, security, and differences in standard operating policies. Although 1,000 doses of MMR vaccine were made available, only about 100 were used.

Analysis of the measles outbreak demonstrates that outbreak response requires not only the state and local health departments but also the schools and healthcare system. Strengthening relationships with these partners before the next crisis could help alleviate the burden of setting up programs when needed. Preparation of a question and answers document for schools and medical providers on outbreak decision making, for instance, could be useful. Specific consideration needs to be given to adults and ancillary staff that work with children in congregate settings like school, who will also need to have their medical records and vaccine status up to date in the event of a future health emergency.

Coordination and communication

Managing an effort as complex and extensive as was required in the measles outbreak requires effective coordination and communication among many individuals and entities. This includes individuals designated as part of the CCPH or DOH incident management teams, as well as many others in CCPH, the DOH, other state and local public agencies (in Oregon as well as Washington), the schools, and the private healthcare system. Communication between emergency management and subject matter experts as well as coordination of the efforts of multiple disciplines including physicians, nurses, epidemiologists, emergency personnel, and educators were also required. Quelling the outbreak in only six generations in about three months is itself evidence that coordination and communication efforts were successful.

Nevertheless, there were challenges during the measles outbreak related to coordination among partners, limited informational sharing, and other aspects of the Incident Command System (ICS) structure that CCPH and the DOH employed. When part of an ICS structure, individuals leave their "day job" and play a role in the IMT. Individuals, however, have existing relationships and practices that they carry from their day jobs. Consequently, there is an inherent tension between functioning within the ICS structure and continuing to maintain external relationships built on collegiality and familiarity as well as professional identification and organizational roles.

For example, the OCIP staff had existing relationships with Clark County healthcare providers through the Vaccine for Children Program. But differences in belief between CCPH and the OCIP about the need for a public vaccination clinic, for instance, might reflect the OCIP practices and relationships with healthcare providers that were not reflected in the ICS response priorities. Similarly, epidemiologists are likely to communicate across jurisdictions when not in ICS. Some saw the "epi-to-epi" calls that were set up to coordinate the efforts in different jurisdictions as a successful way to facilitate active surveillance; others saw these calls as an attempt to circumvent ICS procedures.

The ICS structure is designed to enable rapid coordination and functioning for a jurisdiction so that personnel can effectively operate during the event and return to daily activities as quickly as possible. All deployed staff, regardless of what institution and role they serve in their day job, need to understand and share these goals and basic ICS procedures. For many, however, this was their first response using ICS, and some of the problems experienced during the outbreak were attributed to individuals with a lack of training or understanding of ICS procedures. To address this problem, a summary sheet summarizing the goals of ICS and the roles of players could be prepared to remind personnel what needs to occur now that ICS is activated. The summary should clarify how responsibilities vary during the event. In the initial phase, the focus of the ICS should be on information gathering and understanding what is going on. Later, attention should be focused on operations and communicating with other organizations what needs to be done and managing personnel and other resources that were brought in to address the event. Near the end, the ICS needs to focus on demobilization and setting the stage for organizations to take over seamlessly. The ICS summary should also contain information on optimal communication channels. Additionally, it is also important for all response personnel to understand how to manage relationships outside the ICS structure, as it is unrealistic to think personnel will completely not communicate with colleagues.

In some cases, particular individuals were assigned to the wrong roles in the ICS structure, were not included, or should not have been in the ICS at all. Some of these assignment problems may have been idiosyncratic, but adaptations and modifications to standard ICS procedures for future public health emergencies might be considered. For example, the ICS structure is designed to help local health officers manage an event, not to usurp their authority; as in the CCPH response, neither the health department administrator nor the health officer should be the incident commander due to competing duties and time constraints. Instead, the ICS can support them in decision-making and communicating with the public, the healthcare community, and so on. For the same reason, other subject matter experts might also be excluded from the ICS structure. Another modification to consider is to plan for regular meetings of epidemiologists and other public health experts to facilitate data sharing and expert-to-expert communication necessary for an effective public health response to major infectious disease outbreaks.

Conclusions

Although successful in many ways, the response mounted by CCPH and partners to the 2019 measles outbreak illustrates some important challenges in addressing public health emergencies, as well as strategies to overcome them. To some degree, the challenges reflected specific factors in Clark County, its public health system, and the community in which the first cases emerged. Beyond the specifics, however, this case study illustrates lessons about both public health strategies and emergency management that are potentially applicable to public health emergencies caused by infectious disease outbreaks in other settings.

One challenge was the tension between publicly identifying the Slavic community in which early cases occurred to aid in the outbreak investigation and help people understand what was going on and at the same time avoiding stigmatizing the group and perpetuating a misunderstanding of risk that others faced. Another was responding to false claims on social media, many from outside Clark County, in a way that allowed CCPH to identify some real community concerns and demonstrated that the health department was willing to engage in a dialog with community members who had these concerns. Working now with schools, the medical community, and community groups to build trust and understanding can help improve future emergency responses as well as address vaccine hesitancy in general.

Early emergency declarations allowed CCPH to assemble an incident management team including hundreds of individuals that identified 53 public exposure sites, traced 4,138 contacts of measles cases, made daily monitoring calls to 816. CCPH also worked with schools and child care centers to exclude 849 students in 15 schools in three public school districts and two private schools. As a result, the outbreak was ended in only six generations, or about three months, with only 71 confirmed cases.

Local health departments cannot act alone; they are part of a complex public health emergency management system. Managing this effort required extraordinary coordination and communication among many individuals and entities: the CCPH and DOH incident management teams, many others in CCPH, the DOH, other state and local public agencies, the schools, and the private healthcare system. The response also required communication between emergency management and subject matter experts as well as coordination of the efforts of multiple disciplines including physicians, nurses, epidemiologists, emergency personnel, and educators. Each of these entities has its own responsibilities, authorities, capabilities, and interests. The early implementation of the ICS effectively facilitated much of this coordination and communication, yet there were challenges that will likely need to be addressed in future outbreaks and public health emergencies.

Data sharing

Managing an outbreak requires the sharing of detailed epidemiologic data on potential as well as confirmed cases (sometimes across jurisdictional lines), information on potential exposure sites (including schools), and data on vaccination/immunity status (including for adults in the schools). Although effective *ad hoc* solutions were developed during the outbreak, the development of an outbreak module for the Washington Disease Reporting System and standard approaches to linking school immunization records and the state vaccine registry might facilitate data sharing in future outbreaks.

Multisector collaboration

The 2019 measles outbreak demonstrated, yet again, the critical need for multisector collaboration in dealing with public health emergencies. The response required epidemiologists, vaccine specialists, emergency managers, communication specialists, and many others in CCPH and the DOH; collaboration with the schools and healthcare delivery system; and many others. Each of these disciplines and organizations brings not only different information and capabilities but also different perspectives. Epidemiologists expect to share more detailed case data than the ICS typically includes. The state vaccine program (OCIP) had a different perspective than the local health department on the need for and feasibility of a public vaccine clinic. Multisector collaboration is required to mount the multifactorial response needed, so communication and coordination structures and approaches that recognize the difference in perspective as well as capabilities are required.

Incident Command System

Although the ICS structure proved effective in managing a large multifactorial response, there were challenges related to data sharing and multisectoral collaboration. Communications outside the ICS related to IMT members' day jobs reflected differences in organizational perspectives and created challenges. To some degree, these issues can be addressed through more extensive training about ICS goals and procedures. CCPH and the DOH might also consider, in advance of the next outbreak, the role assignments that are most appropriate for a disease outbreak scenario. Finally, it must be recognized that infectious disease outbreaks are different from other public health emergencies; in particular, they are typically of longer duration and require different approaches at the beginning, middle, and end. While *ad hoc* solutions such as the epi-to-epi calls were implemented during the outbreak were effective, other approaches to expert-to-expert communication and managing relationships outside the ICS appropriate for disease outbreaks might be developed in advance.

Appendix 1. Participants

Clark County Public Health Representatives.
  CCPH Administrator/Health Officer.
  CCPH Community Health & Safety Director.
  CCPH Public Informational Officer.

Region IV Emergency Preparedness Coordinator, CCPH Incident Commander.
Clark Regional Emergency Services Agency & ICS Liaison officer.
CCPH Staff, Emergency Manager of Providence Health during the Incident.
Clark County School System Representatives.
Evergreen Public Schools Public Information Officer.
Evergreen Public Schools Health Services Manager.
Vancouver Public School District, Deputy Superintendent.
Washington State Department of Health Representatives.
Secretary of Health.
Agency Administrator.
Senior Epidemiologist, ICS Subject Matter Expert during the response.
Incident Commander.
Director, Office of Immunizations and Child Profile.
Communications Director.
DOH Representative to local health departments.

Appendix 2. Critical events timeline

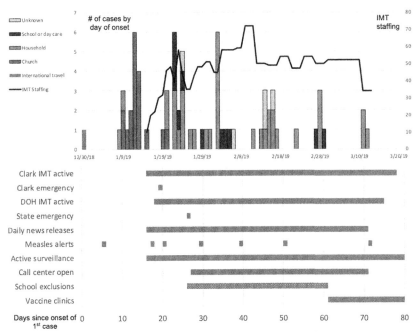

Critical events timeline of measles outbreak in Clark County, Washington.

Critical events timeline of measles outbreak in Clark County, Washington. December 31, 2018 Clark County Public Health (CCPH) notified of a suspected case of measles in a child recently arrived from Ukraine.

January 3, 2019 Measles virus detected in the child's specimens by RT-PCR.

January 15, 2019 First provider health alert & media release sent out, CCPH website and social media updated; Case notification and contact tracing begins.

January 15, 2019 CCPH delegates authority to the Incident Management Team (IMT) to manage the outbreak and IMT is activated.

January 15, 2019 Washington State Department of Health (DOH) IMT activated.

January 18, 2019 Clark County declares a public health emergency.

January 25, 2019 Washington State declares a public health emergency.

January 15, 2019 School exclusions begin.

January 16, 2019 Call center opened.

March 1—8, 2019 Legacy Salmon Creek Medical Center measles vaccines clinics.

March 11, 2019 Call center deactivated; CCPH discontinues daily news releases.

March 12, 2019 DOH IMT deactivated.

March 18, 2019 CCPH IMT deactivated.

April 28, 2019 Outbreak declared over.

References

[1] Papania MJ, Wallace GS, Rota PA, Icenogle JP, Fiebelkorn AP, Armstrong GL, et al. Elimination of endemic measles, rubella, and congenital rubella syndrome from the Western hemisphere: the US experience. JAMA Pediatr 2014;168:148—55.

[2] United States Centers for Disease Control and Prevention. Measles elimination. 2019. www.cdc.gov/measles/elimination.html. [Accessed 27 October 2020].

[3] European Centre for Disease Prevention and Control. Addressing misconceptions on measles vaccination. 2019. www.ecdc.europa.eu/en/measles/prevention-and-control/addressing-misconceptions-measles. [Accessed 27 October 2020].

[4] World Health Organization. Ten threats to global health in 2019. 2019. www.who.int/newsroom/feature-stories/ten-threats-to-global-health-in-2019. [Accessed 27 October 2020].

[5] Washington State Department of Health. Washington state immunization information system (IIS). 2019. waiis.doh.wa.gov/iweb/. [Accessed 27 October 2020].

[6] Hall V, Banerjee E, Kenyon C, Strain A, Griffith J, Como-Sabetti K, et al. Measles outbreak — Minnesota April-May 2017. MMWR 2017;66:713.

[7] Williamson G, Ahmed B, Kumar PS, Ostrov BE, Ericson JE. Vaccine-preventable diseases requiring hospitalization. Pediatrics 2017;140:e20170298.

[8] Patel M, Lee AD, Clemmons NS, Redd SB, Poser S, Blog D, et al. National update on measles cases and outbreaks — United States, January 1-October 1, 2019. MMWR 2019;68:893.

[9] McDonald R, Ruppert PS, Souto M, Johns DE, McKay K, Bessette N, et al. Notes from the field: measles outbreaks from imported cases in Orthodox Jewish communities — New York and New Jersey, 2018-2019. MMWR 2019;68:444.

[10] United States Centers for Disease Control and Prevention. State school and childcare vaccination laws. 2017. www.cdc.gov/phlp/publications/topic/vaccinations.html. [Accessed 27 October 2020].

[11] Melnick A. Vaccinations: science, controversy, and public trust. Washington: Measles and Rubella Initiative; 2019.

[12] California State Senate. California state senate passes Sb 276 to prevent fake medical exemptions that are contributing to measles outbreaks. Sacramento: Government of the State of California; 2019.

[13] New York State Assembly. A02371 summary. Albany: Government of the State of New York; 2019.

[14] National Coalition of State Legislatures. States with religious and philosophical exemptions from school immunization requirements. Washington: National Coalition of State Legislatures; 2019.

[15] Washington State Department of Health. Governor signs law strengthening MMR immunization requirements. Washington State Department of Health; May 10, 2019.

[16] Washington State Department of Health. Washington State Department of Health School Immunization Report. 2019. www.doh.wa.gov/CommunityandEnvironment/Schools/Immunization/SchoolStatusReporting. [Accessed 27 October 2020].

[17] Zucker JR, Rosen JB, Iwamoto M, Arciuolo RJ, Langdon-Embry M, Vora NM, et al. Consequences of undervaccination — measles outbreak, New York City, 2018-2019. N Engl J Med 2020;382:1009—17.

[18] Clark County Government. About Clark county: promising future. 2020. www.clark.wa.gov/county-manager/promising-future. [Accessed 27 October 2020].

[19] United States Census Bureau. QuickFacts Clark county, Washington. 2019. www.census.gov/quickfacts/fact/table/clarkcountywashington,WA/PST045218. [Accessed 27 October 2020].

[20] Washington State Office of the Superintendent. Washington school report card. 2019. https://washingtonstatereportcard.ospi.k12.wa.us/. [Accessed 27 October 2020].

[21] Clark County Government. About public health. 2019. www.clark.wa.gov/public-health/about-public-health. [Accessed 27 October 2020].

[22] Institute of Medicine. Research priorities in emergency preparedness and response for public health systems: a letter report. Washington: The National Academies Press; 2008.

[23] Carlson A, Riethman M, Gastañaduy P, Lee A, Leung J, Holshue M, et al. Notes from the field: community outbreak of measles — Clark county, Washington, 2018-2019. MMWR 2019;68:446.

[24] Aledort JE, Lurie N, Ricci K, Dausey DJ, Stern S. Facilitated look backs: a new quality improvement tool for management of routine annual and pandemic influenza. Santa Monica: RAND Corporation; 2006.

Chapter 4

After-action reviews as a best practice tool for evaluating the response to urban disease outbreaks in Nigeria

Adejare (Jay) Atanda[1], Emmanuel Agogo[2], Kayode Fasominu[3], Folake Lawal[4], Ibrahim Seriki[5], Adesola Ogunsola[6], William Nwachukwu[6], Chioma Dan-Nwafor[6], Oyeladun Okunromade[6], Oyeronke Oyebanji[7], Samuel Mutbam[8], Ifeanyi Okudo[8], Elsie Ilori[6], Chikwe Ihekweazu[7]

[1]*School of Community Health & Policy, Morgan State University, Baltimore, MD, United States;* [2]*Directorate of Prevention Programmes and Knowledge Management, Nigeria Centre for Disease Control, Abuja, Nigeria;* [3]*Volte health Systems, Abuja, Nigeria;* [4]*Medical College of Georgia, Augusta University, Augusta, GA, United States;* [5]*Zamafara State Field Office, World Health Organization, Zamfara, Nigeria;* [6]*Directorate of Surveillance and Epidemiology, Nigeria Centre for Disease Control, Abuja, Nigeria;* [7]*Office of the Director General, Nigeria Centre for Disease Control, Abuja, Nigeria;* [8]*Nigeria Country Office, World Health Organization, Abuja, Nigeria*

In this chapter, we provide evidence of the usefulness of after-action reviews (Aars) as a best practice, frontline tool for evaluating the response to urban disease outbreaks by discussing three Aars conducted in multi-city outbreaks of Lassa Fever (LF), monkeypox (MPX), and cerebrospinal meningitis (CSM). After-action reviews are a qualitative review of actions taken in response to a real-life event, in this case, a public health event, providing a realistic assessment of the ability to implement International Health Regulation (IHR) core capacities. All outbreaks evaluated here were in cities or urban local government areas (LGAs) across more than 20 states in Nigeria, lending credence to the rise of city-states as a locus of risk for infectious disease re-emergence. Aspects of modern life in urban LGAs puts residents at risk for infectious disease outbreaks. For example, encroaching into new environments and destruction of forests for agriculture means that rodents, the reservoir for LF are now in closeness, proximity to humans; as a result, the epidemiology is not only characterized by rodent-to-rodent transmission, but also spillover (i.e., animal-to-human), human-to-human transmission, as well as nosocomial transmission.

Inoculating Cities. https://doi.org/10.1016/B978-0-12-820204-3.00004-8

Regarding MPX, there is no definitive treatment for the disease and gaps remain in our knowledge of the animal reservoir and routes of transmission. These realities highlight the need for rapid identification of the determinants of MPX emergence in Nigeria including the distribution of animal reservoirs and modifications to human behaviors that could have promoted this outbreak. Though the evidence points to secondary human-to-human transmission as common; in Nigeria, primary exposure is thought to stem from direct human contact with reservoirs of infection in exotic rodents or non-human primates during bush meat preparation.

Under the IHR, countries are required to have core capacity for disease surveillance, reporting, notification, verification, response, and collaboration. The Nigeria Centre for Disease Control (NCDC) is recognized by law as the National IHR focal point and coordinates all IHR related activities. The NCDC is committed to IHR implementation at the state- and local-levels; as such, primary response efforts are led by state and local public health officials with central coordination and support provided by NCDC. This includes the provision of guidance, technical and logistic support to the states and local governments for the planning, implementation, and management of diseases of public health importance and on activities to reduce health risk and impact from public health events. NCDC carries out its mandate by working with personnel responsible for executing health security activities at state-local level such as state epidemiologists, and Disease Surveillance and Notification Officers at the LGAs. The NCDC also provides technical support through the deployment of field epidemiologists to support state- and local-level personnel responding to public health events. To foster cross-learning by LGA staffers from different urbanicity levels and from across multiple states, the NCDC hosts an annual critical review meeting of epidemiologists where previous outbreak responses are reviewed, and lessons learned are shared. This platform was leveraged to conduct these AARs. Using an approach that combined self-evaluation and peer review, NCDC convened key actors between July and August 2018 to review the response experience and assess preparedness for future outbreaks.

For the three outbreaks, the focus was to evaluate outbreak preparedness and response, identifying best practices and challenges and drawing lessons for improved response to future disease outbreaks. Furthermore, the AARs provided an opportunity to review challenges faced by technical areas during the response phase, and those that will require frequent and deliberate improvement during "peacetime" because of their importance to the overall success of subsequent response efforts. The AARs determined how actions were implemented as opposed to how they were planned, identified processes that worked and those that didn't, determined root causes for success and failure, and finally, provided SMART (i.e., Specific, Measurable, Achievable, Relevant, and Time-bound) corrective recommendations to improve future actions and to strengthen cities' pandemic readiness at the local level.

A brief history of after-action reviews

After-action reviews were not routine until about three decades ago. The earliest record of use and origin of AARs appear to be from the United States (US) Army. Historian S.L.A. Marshall documented events during World War II, the Korean and Vietnam Wars, and this is thought to be the genesis of the modern AAR [1]. Marshall spoke with soldiers in theater, immediately after combat actions, and although sometimes criticized because of his small samples, his efforts were arguably the first AARs. A lack of modern technology may have played a significant role, restricting first-hand observations and documentation of battles, as reporters were limited by place and time. Marshall's efforts were an attempt to describe what actually happened, relying on the real time, post-event memories of participants.

Other accounts emphasize the role of the US Army in developing and institutionalizing the process of conducting AARs that has since cascaded down to the Navy, Air Force, and Marines [2,3]. The AAR has evolved over time and now serves as the military's primary feedback process after collective training events as required by regulation. When done properly, AARs are a valuable tool to help units understand their most recent performance - whether a training event or a combat activity [4]. As such, guidance was authored for implementation across the various military units [3]. Repeatedly, this guidance stresses that an AAR is neither an evaluation, nor a critique, and that leaders must not lecture their units or interject personal opinions. It has since evolved following its spread throughout the military, and in slightly altered form, to the corporate and business community. It has also taken a foothold in public health and been adapted accordingly.

The definition and purpose of AARs in the military is stated as [4];

A professional discussion of an event, focused on performance standards, that enables soldiers to discover for themselves what happened, why it happened, and how to sustain strengths and improve on weaknesses. It is a tool leaders and units can use to get maximum benefit from every mission or task. It provides - candid insights into specific soldier, leader, and unit strengths and weaknesses from various perspectives, feedback and insight critical to battle-focused training, and details often lacking in evaluation reports alone.

World Health Organization recommended after-action reviews formats for public health events

There are four recommended formats for AARs. Three of these use a single format based on the number and location of participants, the complexity of the event, and the number of technical areas to be reviewed. They include the (1) debrief, (2) key informant interviews, and (3) working group format. The fourth method combines two or more of the other three methods. Accordingly, the choice of format depends largely on the number of technical areas to be evaluated, the complexity of the event, the location and proximity of participants, and the resources available. Regardless of the format selected, all require the AAR to be conducted within three months following the declaration of an outbreak being over by a state ministry of health.

Debriefing is the simplest of the formats. It is an informal, in-depth review of one technical area done in a facilitated small group discussion with a narrow scope for a small event or response. Key informant interviews, on the other hand, are used when participants are not in the same location and cannot be readily brought together. This format is often used to review multiple technical areas where key stakeholders are interviewed in one-on-one sessions, and summaries are then compiled into a single report. Importantly, this format allows for more personal and anonymous expressions of concerns than in group settings, however, because the report is not collectively discussed by the group, participants may not take full ownership. Finally, the working group format is used for the extensive review of multiple technical areas when all stakeholders can be present in the same location for parallel group discussions. This format allows for cross-review of each technical group and areas of overlap, cross-function, and transitions between the technical areas.

The NCDC employed the working group format for the three AARs examined here, capitalizing on the ability to bring relevant stakeholders to a single meeting point and examined multiple technical areas during working group discussions. The NCDC followed standardized World Health Organization (WHO) methodology by examining five key areas: (1) what was in place prior to the event, (2) the timeline of key events, (3) what went well or did not go well during the event, (4) areas needing improvements, and (5) recommendations for improvement. This was done across five incident management structure (IMS) core pillars for outbreak response (Box 4.1). These pillars are broad technical categories that allowed for the combination of several specific technical areas or functions and were used to structure the conduct of the AARs.

For guidance, NCDC used the WHO AAR toolkit that included the guidance for AARs and working group facilitators and participants manuals.

BOX 4.1 Core pillars of incident management structures for outbreak response

Coordination and logistics.
 Case management and infection prevention and control.
 Laboratory capacity.
 Risk communication and social mobilization.
 Disease surveillance.

A pre-workshop meeting was held for facilitators, notetakers, and report writers to familiarize them with the tools, tasks, and expectations for the workshop. From July through August 2018, the NCDC brought together multiple stakeholders for the AARs despite the primary locations of actors being different states and cities around the country. Each AAR meeting started with a general presentation of the objectives of the AAR. Participants were then divided into technical groups, aligned with IMS response pillar areas. Each group revisited their relevant timelines, examined best practices, and challenges using the "5 Whys" technique of root cause analysis and made recommendations for future action. An additional step titled "wall café" was included for the CSM and MPX AARs where participants in each technical group visited other groups to assess their progress and offer suggestions. The findings from each group were then presented to all participants by a group representative, discussed and recommendations examined before adoption and collation by report writers. Following the conduct of the AAR, plans and recommendations developed by participants and stakeholders who were from national, sub-national and across ministries, departments and agencies (MDAs) were used to inform subsequent outbreak response with coordination led by NCDC as the national IHR focal point.

After-action reviews in public health emergency response

The public health community — by virtue of often working with the military in humanitarian crisis responses — has adopted AARs to improve organizational learning [2,5]. The WHO, in response to an increasing threat of disease outbreaks and their international spread, introduced the revised IHR in 2005 that entered into force in 2007. Under the IHR, State Parties are obliged to develop and maintain minimum core capacities for surveillance and response, in order to detect, assess, notify, and respond to any potential public health event of international concern ensuring mutual accountability for health security.

Complying with the recommendations of the IHR begins with the mandatory State Parties self-assessment annual reporting (SPAR), and the voluntary AARs, simulation exercises, and external evaluation processes using the Joint External Evaluation (JEE) tool. The JEE and SPAR assess national preparedness capacities and provides a comprehensive picture of the capacity and preparedness of Member States in the implementation of the 19 IHR core capacities as part of a five-year global strategic plan to improve public health preparedness and response (Fig. 4.1) [6−9]. Within the context of preparedness and as a tool for evaluation of responses to outbreaks (Fig. 4.2), AARs provide a means to observe how well preparedness systems perform in real-world conditions and can help to identify and address gaps in national and global public health emergency preparedness systems, besides monitoring compliance with other IHR requirements [10]. They are also applied to ensure that plans, processes, and other capacities are up to date and to make the best

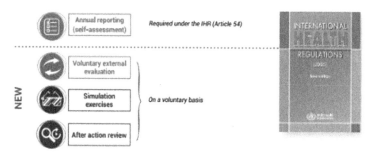

FIG. 4.1 International Health Regulations Monitoring and evaluation framework [19].

possible use of limited resources. Similar to the guidance provided by the US Army, the WHO has recently published a country implementation guidance document, "After action reviews and simulation exercises: under the International Health Regulations 2005 monitoring and evaluation framework" to ensure the integrity of the process for conducting AARs. The WHO guidance describes an AAR as;

A means of identifying and documenting best practices and challenges demonstrated by the response to the event. An AAR seeks to identify actions that need to be implemented immediately, to ensure better preparation for the next event; medium- and long-term actions needed to strengthen and institutionalize the necessary capabilities of the public health system.

FIG. 4.2 After action reviews in preparedness and response cycle [19].

Additionally,

An AAR is designed to be flexible. It can be adapted to fit the event under review, and the organization and systems involved. Its success hinges on the ability to bring relevant response stakeholders together in an environment where they can analyze actions taken during the response in a critical and systematic fashion and identify areas for improvement.

An AAR aims to assess the functional capacity of existing systems, identify challenges and best practices encountered during a response, document and share these experiences with other relevant stakeholders, identify practical actions for improving existing capacities and capitalizing on best practices, and improve preparedness, readiness and response plans (Box 4.2). After conducting an AAR, consensus will also need to be reached on how widely results will be shared in order to codify lessons, experiences, examples, and models; advocate for support for preparedness and readiness actions.

General characteristics and epidemiology of the urban outbreaks

All outbreaks evaluated in the three AARs occurred across more than 20 city-states and towns across Nigeria. The CSM outbreak occurred in urban centers across 14 states and commenced with an index case identified in epidemiological (epi) week 47, 2017; peaked and ended in epi week 15 and 18 respectively of 2018; and resulted in 3467 suspected cases with a case fatality rate (CFR) of 8.7%. The LF outbreak commenced with an index case on January 14, 2018 with cases reported in urban centers across 21 states. By epi week 22 of 2018, there were 1982 suspected LF cases, with 442 laboratory cases and a CFR of 25%. The MPX outbreak commenced in September 2017 with an index case identified as a boy in contact with a neighbor's pet monkey in Agbura, a large town on the outskirts of Yenagoa, Bayelsa's state capital, in the Niger-Delta. The outbreak resulted in 225 cases with a CFR of 6%. This was the first documented MPX outbreak since three cases were identified in 1971, and genomic sequencing links the virus strains from both outbreaks. Epidemic curves for the three outbreaks are below (Fig. 4.3).

BOX 4.2 Aims and objectives of health security after-action reviews

Assess the functional capacity of existing systems to prepare, prevent, detect and respond to a public health event;
  Identify challenges and best practices encountered during the response;
  Document and share experiences of response with stakeholders;
  Identify practical actions for improving existing capacities and capitalizing on best practices;
  Improve preparedness, readiness and response plans.

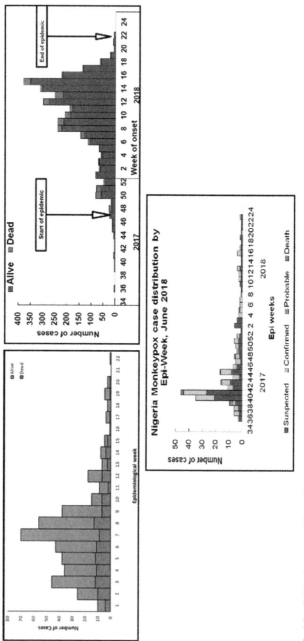

FIG. 4.3 Epidemic curve of the outbreaks (1) Lassa fever cases January–June 2018, (2) CSM cases 2017/2018, and (3) Monkeypox cases 2017/2018.

After-action reviews in outbreak response in Nigerian cities

Nigeria is a federal republic with a presidential constitutional system modeled after the US but with Westminster model system influences. The constitution provides for a separation of powers among the three tiers of government — federal, state, and local. Each of the 36 "politically autonomous states" in Nigeria has its own budget, priorities, and constitutional authority for health sector interventions including public health emergencies [11]. Local government areas are administered by a council consisting of a chairman as chief executive and elected members referred to as councilors [11]. Each of Nigeria's 774 LGAs is made up of contiguous towns and cities of varying urbanicity levels, villages etc. and further subdivided into 10–15 wards. Ward residents play a direct and full part in administration through their elected representatives who exercise power or undertake functions under the general authority of the national government such as providing a limited range of social amenities. The councils are also vested with the responsibility of health services delivery, particularly primary health care, thus enhancing the grassroots healthcare delivery system [11]. Devolving provision of services and development activities at this level ensures they are responsive to local wishes. However, LGAs, their wards, and local-level leadership still play a subordinate role to state governments because they have no separate source of funding and allocated federal funds is disbursed through state governments [12]. The NCDC is the country's national public health institute (and national IHR focal point) with the mandate to lead the prevention, preparedness for, detection of, and response to infectious disease outbreaks. Article N, Part II of the NCDC Act (2018) mandates the agency to;

> *Provide guidance, technical and logistic support to the States and Local Governments for the planning, implementation and management of diseases of public health importance and on activities to reduce health risk and impact from public health events.*

Following the guidance from the WHO, AARs are now commonly conducted following major emergency response activities across the world including Nigeria [13,14]. The NCDC, as the national IHR focal point, worked closely with the WHO to conduct three AARs post-response to multi-city outbreaks of MPX, LF, and Cerebrospinal Meningitis (CSM) in 2018 [15–18]. While it is best practice to conduct AARs as part of routine emergency management procedures for all emergency responses, resources and time constraints make this a lofty aspiration. As such, the WHO provides guidance for selecting events for an AAR when one or more of predefined characteristics are met (Box 4.3) [19,20].

The three AARs conducted in Nigeria were for outbreaks that met a majority of the outlined characteristics. For example, the MPX outbreak was the

---

BOX 4.3 Event characteristics that warrant an after-action review according to guidance from the World Health Organization [19,20]

At least one of the 13 core capacities as defined under the SPAR was tested by the event;

The event was declared as a public health emergency of international concern;

The WHO was notified of the event as required under IHR regulations;

The event was a graded emergency under the WHO Emergency Response Framework as a level two or three event;

The public health emergency operations center was activated;

The event involved coordination and collaboration with sectors that do not routinely collaborate;

The AAR was recommended by WHO following an event that constituted an opportunity for collective learning and performance improvement.

---

first outbreak of that particular disease in nearly 40 years and resulted in travel-associated importation to the United Kingdom and Israel, infection of healthcare workers, and cases in a prison facility [17,21]. The LF outbreak was the largest on record by the time the AAR was conducted with the confirmed cases reported by mid-2018 exceeding the total reported in 2017 and included many healthcare worker infections [22]. Finally, the CSM outbreak was unprecedented as it was largely caused by the type C strain of the disease-causing bacteria *Neisseria meningitidis* [14].

Key findings: best practices and recommendations for urban environments

Upon review, several themes emerged from the three urban AARs that could serve as best practices for pandemic preparedness in urban environments.

Standard operating procedures, guidelines, and response plans

This theme concerns a roadmap for responders and is reflected across the five core pillars of outbreak response in the IMS. As part of coordination, for instance, partners in states stated existing CSM and LF guidelines and standard operating procedures (SOPs) provided by national authorities helped with effective case management. Also highlighted, was how the process of national staff helping state partners to develop a response plan encouraged local stakeholder buy-in. Strict adherence to SOPs for sample collection and transportation resulted in prompt sample reception for analysis and this in turn resulted in timely patient management. Good infection prevention and control (IPC) practices were especially highlighted for the MPX response where no known treatment is available and standard precautions were required to be strongly implemented. Standardized case definitions were also shared with health facilities, communities, and other partners. In addition, the use of standardized definitions for alert and epidemic thresholds helped with identifying the beginning and end of outbreaks. The availability of a risk communication plan also helped to ensure SMART implementation.

The importance of partnerships and collaborations

Partnerships and collaborations (i.e., inter-sectoral, multisectoral, and cross-border) was another theme identified through the thematic analysis. This theme is the key driver for the coordination and logistics IMS pillar. For example, public-private partnerships (PPPs) between the NCDC and TRANEX — a logistics company engaged in domestic and international express delivery, haulage, and freight — provided rapid laboratory sample transportation for testing. Collaboration between states, non-government organizations, and other partners helped ensure a larger community reach during awareness and sensitization for LF. Participants described inter-sectoral, multisectoral, and cross-border collaborations that helped with the timely detection of imported cases and to prevent spread across sub-national borders. Coordination at the ward level helped with the timely detection of local cases. Coordination between states and national also ensured rapid sample referral to the NCDC national reference laboratory for analysis when needed. There was also engagement of other relevant institutions, such as religious and traditional rulers in sensitization efforts, particularly for the response to the LF outbreak. In addition, highly placed government officials and political officials acted as champions for LF sensitization. Community observer groups were also formed to ensure adherence to food safety measures such as environmental sanitation and food handling during LF outbreaks. These, plus the existence of an incident command system ensured a good working relationship between states and NCDC. Periodic outbreak review meetings at the state- and national-levels were enabled by strong public health leadership. Mirroring PPPs is the synergy between state and federal agencies, and intra-federal or state agency collaborations. For example, weather forecasts and climate products from the Nigerian Meteorological Agency were used by NCDC to predict, identify, and plan for at-risk states.

Quality control in laboratory practices

Besides prompt sample reception for analysis resulting from strict adherence to laboratory SOPs, the preservation of isolates for internal quality control was identified as a best practice. Participants described the availability and right application of sample collection protocols, mandating use of personal protective equipment (PPE), and guiding sample collection and transportation depending on sample type — this was specifically noted for MPX because of its novel nature and lack of treatment or vaccine. Quality control measures like these were thought to speed up the analysis, release, and sharing of laboratory results, ensuring timely patient management.

The vaccination of laboratory staff involved in sample handling and analysis as a biosafety measure was highlighted for CSM. The development of a standard, national testing algorithm for LF also ensured consistent application across laboratories. Additionally, participants highlighted the importance of monitoring and supervision of vaccination teams to maintain the integrity of the vaccine cold chain as part of the coordination/logistics IMS pillar for CSM outbreak response.

Supporting systems and tools

This theme is concerned with systems and tools that support the preparedness, response, or recovery effort. This includes the formation and timely activation of local emergency operation centers with the support of states — that built on the existence of an incident command system, the generation of situation reports and press releases, availability of a functional indicator and event-based surveillance system at the state and national level. A real-time surveillance data platform, the Surveillance Outbreak Response Management and Analysis System, was also deployed to some states for case-based surveillance. States and local communities benefited from enhanced community-based surveillance where focal points for surveillance were identified in health facilities and communities. These focal points served as contact points for the distribution of case definitions and helped identify established surveillance systems locally for integration with the national system.

Data analytics was an important skillset identified — surveillance and epidemiology data were analyzed weekly at the national and state levels. GIS analytic capabilities helped with the mapping of locations of laboratory testing facilities. It was also important to think strategically about supporting providers with case management and IPC. For example, shared spaces by laboratory services for supportive investigations in the isolation/treatment centers and intensive care units, was identified as a best practice.

One important and recurring subtheme under the coordination and logistics IMS pillar was the availability of funds and their timely release, for example, to procure response materials and drugs. Participants described "the availability of logistics" — drugs, consumables, and vehicles. In addition, participants noted that providing cash stipends to LF case management teams was motivation to quickly complete important outbreak response tasks.

Key findings: limitations and challenges

States' responsibilities for health in alignment with Nigeria's governance structure, include the preparedness for and response to disease outbreaks within their boundaries. The WHO Integrated Disease Surveillance and Response Technical Guidelines call for the establishment of Epidemic Preparedness and Response or Public Health Emergency Management Committees at state- and local-levels, supported by NCDC. These committees lead the coordination of activities related to emergency preparedness and response as part of their terms of reference. Members of the committee should include relevant personnel from all MDAs with activities and a mandate to protect their citizens from public health threats. Such multisectoral committees are also important in domesticating and ensuring the full implementation of recommendations within states following the AARs. These committees, however, have not functioned as envisioned across the various administrative levels and recognizing this was a watershed moment consequential to the conduct of the AARs for these outbreaks.

The impact of the non-existent or partially constituted committees often results in the lack of follow-through with recommendations from the AARs — even when recommendations are SMART — resulting in the partial implementation of preparedness and response plans and the prolongation of outbreaks. Limited human expertize and material resources and inadequate funding, likewise, impede the ability of committees to galvanize necessary support at the state or local levels, or leverage existing policies and platforms to embed these new recommendations for implementation. Implementation of recommendations from AARs are also faced with the red tape or bureaucratic processes common within government agencies and duplication of efforts across MDAs.

Inadequate funding, for example, was a recurring theme for all three AARs in the coordination and logistics pillar that resulted in a delayed or poor outbreak response. The lack of a dedicated budget-line for risk communication activities meant that health education officers were poorly remunerated, in turn resulting in turnover without replacement and inadequate staff to translate materials into local languages. Health education officers also did not participate in outbreak response or meet regularly at the LGAs, so training did not cascade down from the national to grassroots level. Funding challenges also besieged the surveillance IMS pillar, especially for training and capacity building. This trickled down to the inability to use data and reporting tools when available.

For re-emerging infections like MPX, SOPs, guidelines, and treatment protocols were unavailable at the early stages of the response and there was poor knowledge of the case definition. Under the case management/IPC pillar, a common theme was substandard care resulting in poor treatment outcomes and high case fatality and morbidity. This was due to several factors including poor support for the referral of cases, the use of substandard drugs, high workloads for healthcare providers, and inadequate medical supplies (e.g., stockout of consumables such as PPE and medications such as Ribavirin for management of LF, vaccines for CSM, etc.). The laboratory IMS pillar faced challenges such as insufficient testing kits and lack of culture facilities in some urban health centers. When available, culture media had a high contamination rate and laboratory reagents deteriorated due to power outages or inadequate storage facilities for samples and isolates, resulting in false, negative results. Finally, few laboratories were available to test LF samples and these were easily overwhelmed, and it was also difficult to harmonize data generated in the laboratory with surveillance data.

Conclusion

The AARs conducted in Nigeria provided an opportunity to evaluate the responses during these outbreaks in urban settings. Within the NCDC, the use of AARs are now embedded within the institute's monitoring and evaluation framework and fits into the annual strategy of the Health Emergency Preparedness and Response directorate. These AARs are carried out continuously and conducted following responses to outbreaks with coordination and recommendations followed through by the emergency response department at the NCDC.

The five pillars helped to concretize the discussions and eased the development of implementation plans, the assigning of roles, and the tracking of expected deliverables. The AARs were also conducted within three months following the declaration of an outbreak being over as stipulated in the WHO guidance on AARs. This ensured that a majority of the stakeholders were still capable of contributing meaningfully to the discussions while limiting recall bias. It also ensured the collective momentum generated during the response phase had not waned and was channeled to ensure better plans are in place for future responses to infectious disease outbreaks. Though voluntary, AARs have helped Nigeria assess its operational capability for public health preparedness and response. This is important given findings from JEE reports conducted in 2017 where gaps were identified across various IHR core capacities.

In Nigeria, the AAR process further emphasized the importance of urban LGAs/city-states as a key first-line of response to infectious disease outbreaks. Majority of the outbreaks occurred in cities and urban centers that are densely populated and hubs for commerce. Historically, in Nigeria, coastal cities have acted as the primary point of diffusion of pandemics, followed by spread inland. This was the case with the 1918 influenza pandemic that arrived on the shores of Lagos, capital of the colony and protectorate of Nigeria, in September 1918 [23]. It spread further inland along railway transportation routes to cities like Abeokuta, Ibadan; and as far as Onitsha and Calabar in the southeast; and Bornu, Kano and Sokoto in the northeast, and northwest respectively. Despite strict quarantine and isolation measures, the disease spread killing an estimated 500,000 Nigerians across more than 20 towns in 6 months [23]. Today similar trends exist. Infectious diseases are more likely to be introduced through air travel and diffusion inland via land transportation routes, although it is now in privately owned commuter buses and cars where the government has little or no oversight, as opposed to railway routes.

The primary purpose of any AAR is to identify gaps, strengths, best practices, and lessons, so that procedures for improvement can then be implemented at the necessary administrative levels of government, in this case LGAs. These outcomes must be captured and reviewed in a structured, timely manner in order to ensure that all those affected can benefit and that improvements are made. After-action review findings can reassure the public health community that commitment to IHR is strong and that measures are being taken to address identified gaps. Further, the sharing of lessons identified may benefit other countries facing similar challenges or hazards and may provide a basis for follow-on support and funding allocation. In light of recent outbreaks, state and local authorities want to avoid repeating past mistakes and have shared with one another successful key approaches in the deployment of containment strategies such as contact tracing, tracking possible vectors, curfews, shutdowns and statewide lockdowns. Communication strategies and enlightenment campaigns serve a key role in the development of trust, an immeasurable but fundamental public health resource. Specific campaigns, usually radio jingles, are often shared and readapted by localities as part of disease mitigation strategies stressing social distancing, pop-up hand washing stations in public spaces such as markets, schools and even encouraging alternative meat sources outside of "bush meat" consumption suspected to drive some disease outbreaks.

More recent shared lessons include the adaptive or flexible reuse of public facilities such as football stadiums as testing centers or even for the deployment of temporary isolation centers and field hospitals first implemented in Lagos as a PPP with the Lagos chapter of the Young Presidents' Organization, a global leadership community of chief executives [24]. The Lagos experience served as a good model for other Nigerian cities starting field hospitals and showed that a common or shared vision strengthens community relationships even during a pandemic where social distancing measures that weaken social networks are in force.

The AARs also revealed disparities between cities in terms of available resources to prepare, prevent, detect, and respond to a public health event. This manifested in raw financial terms — the level of economic activities in a locality determined the level of financial resources available to health authorities. Another was the availability of trained healthcare workers who usually cluster in large, urban cities. The AARs were an opportunity for staff from different localities to start collaborations where those with more resources helped provide training or even deployed professionals with specialized expertize (e.g., epidemiologists to those cities with less resources).

Even within urban centers, pockets of inequality exist, for example, a hyperconnected and major air transition hub like Lagos still contains informal settlements like the less economically developed and growing, Makoko-Iwaya waterfront community, but lacking innovations like sewers or water filtration systems, making them potential disease hotspots that can amplify pandemic risk. In other urban centers like Ibadan city, the history of disease spread conforms to the generalization that disease diffusion is downward in the urban hierarchy from higher-ordered centers (cities and towns) to lower-ordered centers (villages) [25]. At least nine towns had cholera by the time the first villages reported an infection during the 1971 outbreak and similar to the 1918 influenza pandemic, transportation routes formed the channels of diffusion [25].

The recommendations provided during the AARs are a useful resource for the development of local action plans for health security by cities, that build on the JEE reports, and the National Action Plan for Health Security. Implementation of the AAR recommendations at the city-LGA level will require investments in sanitary infrastructure, help strengthen capacity, and ensure their operational readiness for future public health threats and events.

References

[1] Morrison JE, Meliza LL. Foundations of the after-action review process. Fort Belvoir: United States Army Research Institute for the Behavioral and Social Sciences; 1999.

[2] Stoto MA, Nelson C, Piltch-Loeb R, Mayigane LN, Copper F, Chungong S. Getting the most from after action reviews to improve global health security. Glob Health 2019;15:58—68.

[3] Salter MS, Klein GE. After action reviews: current observations and recommendations. Vienna: Wexford Group International Inc; 2007.

[4] Department of the Army. Training circular 25-20: a leader's guide to after-action reviews. Washington: Department of the Army; 1993.

[5] Savoia E, Agboola F, Biddinger PD. Use of after-action reports (AARs) to promote organizational and systems learning in emergency preparedness. Int J Environ Res Publ Health 2012;9:2949—63.

[6] World Health Organization. Implementation of international health regulations. New Delhi: WHO Regional Office for South-East Asia; 2005. 2008.

[7] World Health Organization. Strengthening response to pandemics and other public-health emergencies: report of the review committee on the functioning of the international health regulations (2005) and on pandemic influenza (H1N1) 2009. Geneva: WHO; 2011.

[8] Talisuna A, Yahaya AA, Rajatonirina SC, Stephen M, Oke A, Mpairwe A, et al. Joint external evaluation of the International Health Regulation (2005) capacities: current status and lessons learnt in the WHO African region. BMJ Glob Health 2019;4:e001312.

[9] Bell E, Tappero JW, Ijaz K, Bartee M, Fernandez J, Burris H, et al. Joint External Evaluation—development and scale-up of global multisectoral health capacity evaluation process. Emerg Infect Dis 2017;23:S33—39.

[10] Piltch-Loeb R, Kraemer JD, Nelson C, Stoto MA. A public health emergency preparedness critical incident registry. Biosecur Bioterrorism 2014;12:132—43.

[11]   Adeyemo D. Local government autonomy in Nigeria: a historical perspective. J Soc Sci 2005;10:77−87.

[12]   Awotokun K. Local government administration under 1999 constitution in Nigeria. J Soc Sci 2005;10:129−34.

[13]   Okunromade OF, Lokossou VK, Anya I, Dada AO, Njidda AM, Disu YO, et al. Performance of the public health system during a full-scale yellow fever simulation exercise in Lagos State, Nigeria, in 2018: how prepared are we for the next outbreak? Health Secur 2019;17:485−94.

[14]   Strategic Partnership for IHR and Health Security. After action review (AAR) on Lassa fever, Abuja, federal capital territory. Nigeria: World Health Organization; June 4−7, 2018.

[15]   Akpede GO, Asogun DA, Okogbenin SA, Okokhere PO. Lassa fever outbreaks in Nigeria. Expert Rev Anti-infect Ther 2018;16:663−6.

[16]   Nnadi C, Oladejo J, Yennan S, Ogunleye A, Agbai C, Bakare L, et al. Large outbreak of Neisseria meningitidis serogroup C−Nigeria, December 2016−June 2017. MMWR 2017;66:1352.

[17]   Yinka-Ogunleye A, Aruna O, Dalhat M, Ogoina D, McCollum A, Disu Y, et al. Outbreak of human monkeypox in Nigeria in 2017−18: a clinical and epidemiological report. Lancet Infect Dis 2019;19:872−9.

[18]   Eteng W-E, Mandra A, Doty J, Yinka-Ogunleye A, Aruna S, Reynolds MG, et al. Notes from the field: responding to an outbreak of monkeypox using the one health approach − Nigeria, 2017−2018. MMWR 2018;67:1040.

[19]   World Health Organization. Guidance for after action review (AAR). Geneva: WHO; 2019.

[20]   World Health Organization. After action reviews and simulation exercises: under the international health regulations 2005 monitoring and evaluation framework (IHR MEF). Geneva: WHO; 2019.

[21]   Kabuga AI, El Zowalaty ME. A review of the monkeypox virus and a recent outbreak of skin rash disease in Nigeria. J Med Virol 2019;91:533−40.

[22]   The Nigeria Center for Disease Control (NCDC). 2018 Lassa fever outbreak response after action review meeting. Lagos: NCDC; 2018.

[23]   Ohadike DC. Diffusion and physiological responses to the influenza pandemic of 1918−19 in Nigeria. Soc Sci Med 1991;32. 1393−39.

[24]   Tassabehji R. Lagos Young Presidents organization community steps up to build COVID-19 field hospital. Young Presidents Organization; May 11, 2020.

[25]   Adesina H. The diffusion of cholera outside Ibadan City, Nigeria, 1971. Soc Sci Med 1984;18:421−8.

Chapter 5

Developing a more effective locally led response to the HIV epidemic in Blantyre, Malawi

Anna M. Carter[1], Chimwemwe Mablekisi[2], Gift Kawalazira[3], Tyler R. Smith[4], Sara M. Allinder[1], Charles B. Holmes[1]

[1]*Center for Innovation in Global Health, Georgetown University, Washington, DC, United States;* [2]*The Malawi National AIDS Commission, Lilongwe, Malawi;* [3]*Blantyre District Health Office, Government of Malawi, Blantyre, Malawi;* [4]*Cooper/Smith, Washington, DC, United States*

Malawi has led a highly successful human immunodeficiency virus (HIV) treatment response that has brought the country close to achieving international goals regarding knowledge of infection status, treatment, and viral suppression. However, new HIV infections persist and the country has much work to do to reduce the rate of infection. Accordingly, a new approach for HIV prevention is needed to accelerate the decline in new HIV infections and sustain epidemic control.

The city of Blantyre is the oldest urban center in Malawi and represents the commercial and industrial capital of the country — attracting many young people and migrant workers to the city in search of economic opportunities. Still, within the city, formal employment rates remain low and poverty rates remain high. This has resulted in a unique local-level HIV risk environment. As a result of this, the city's HIV burden is much higher than that experienced elsewhere in the country. This chapter details an innovative effort to enhance Blantyre's HIV prevention programs through the development of a multi-sectoral, district-level system for reducing the transmission and risk of HIV at the local-level.

HIV/AIDS in Malawi

Malawi has led a highly successful HIV treatment response that has brought the country close to achieving the Joint United Nations Programme on HIV and AIDS (UNAIDS) goals of 90% of people living with HIV know their status, 90% of all diagnosed infections on treatment, and 90% of all people

Inoculating Cities. https://doi.org/10.1016/B978-0-12-820204-3.00005-X
79

on treatment are virally suppressed (Malawi achieved 90-79-72 in 2019) [1]. Still, like other countries in the region, Malawi is not on track to reach the UNAIDS goal of a 70% reduction in new HIV infections by 2020, a situation further exacerbated by the COVID-19 pandemic, which reduced access to health information and services throughout the country. In 2019, 33,000 new HIV infections were recorded in the country, with a significant proportion occurring among young people aged 15–24 years [1]. The ongoing high levels of HIV infection, along with the country's expansive population (Fig. 5.1), and a fragmented HIV prevention response present a grave risk to addressing persistent reservoirs of infection and controlling the HIV epidemic over the long-term.

The Government of Malawi has recognized the need for a comprehensive HIV prevention plan to address these challenges. The National HIV Prevention Strategy (2018–20), based on the UNAIDS Prevention Coalition's

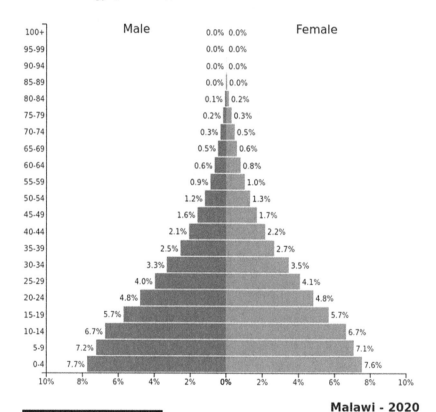

Malawi - 2020
Population: **19,129,955**

PopulationPyramid.net

FIG. 5.1 Malawi population pyramid, 2020. © *December 2019 by PopulationPyramid.net, made available under a Creative Commons license CC BY 3.0 IGO: http://creativecommons.org/ licenses/by/3.0/igo/.*

five pillars − voluntary medical male circumcision (VMMC), condoms, pre-exposure prophylaxis (PrEP), combination prevention for adolescent girls and young women and key populations such as female sex workers, men who have sex with men and injectable drug users, and treatment as prevention − the strategy sets ambitious goals to reduce annual new infections by approximately two-thirds [2,3]. Still, Malawi's uptake of key primary prevention interventions has been limited, with only 28% achievement of VMMC targets, 53% condom use among non-regular partners in 15−24-year-old persons, and limited multi-sectoral programming for adolescent girls and young women [3].

A new approach for HIV prevention is needed to accelerate the decline in new HIV infections and sustain epidemic control. To enhance its HIV prevention programs, the government of Malawi with the National AIDS Commission enlisted the support of a consortium of partners led by Georgetown University's Center for Innovation in Global Health to catalyze the development of an innovative, district-level system for the sustained prevention of HIV infection that is fully embedded in local government, civil society, and multi-sectoral structures. The five-year effort takes a health systems approach to strengthen and institutionalize an innovative and data-driven HIV prevention delivery system at the district-level that is equipped to detect and target risk, generate demand for HIV prevention services, effectively deliver prevention products and interventions, and enable effective utilization of prevention interventions during periods of risk by addressing structural challenges. The effort supports core capabilities required to underpin optimized functioning of the HIV prevention cascade [4], embedded within local systems, in order to provide the government of Malawi, civil society, and other partners with information to improve the HIV prevention response and replicate the approach throughout the country and region.

The southern district of Blantyre was chosen as the initial district for implementation because of its high rates of HIV incidence and troubling structural environment. Blantyre's HIV burden is nearly twice that of the national average, with 18% of Blantyre's population living with HIV [5]. Blantyre district has a population of 1.4 million people of which approximately 800,000 live in the urban center [6]. Blantyre is the oldest urban center in Malawi and operates as the country's commercial and industrial capital, attracting many young people and migrant workers to the city in search of economic opportunities. Yet, formal employment rates remain low and poverty rates remain high, resulting in an acute structural risk environment for HIV acquisition characterized by high rates of gender-based violence and survival and transactional sex.

Many non-governmental organizations, private service providers, and bilateral and multi-lateral organizations and their implementing partners are working in Blantyre to address these challenges and scale-up primary HIV prevention services. The United States' President's Emergency Plan for AIDS

Relief (PEPFAR) and the Global Fund to Fight AIDS, Tuberculosis and Malaria have invested significant resources in the region and have contributed to early successes of testing, treatment, and prevention targets. Blantyre is one of several priority scale-up districts for PEPFAR programming which has led to increased investment in surveillance and testing strategies, VMMC provision, and comprehensive programming for populations at increased risk, such as adolescent girls and young women [7]. These activities are implemented by a number of local non-governmental, faith-based, and community-based organizations and international partner affiliates across the district. Blantyre, like many districts, lacks the systems and resources needed to coordinate the range of activities and diverse set of actors in order to maximize long-term impact across investments.

Governance of multiple sectors in Blantyre remains in transition due to an incomplete devolution process and unclear responsibilities between the two leading local government entities — the Blantyre District Council and the Blantyre City Council. Responsibilities for health are shared between city and district governments, with the District residing over rural areas and the city responsible for service delivery within urban areas. However, actual day-to-day responsibilities for health service delivery and funding lines reside largely with the District. In addition to the Blantyre City and District Councils, there are eight Traditional Authorities in Blantyre. These informal bodies have played a continuous role in local governance and provide critical links to communities, especially in rural areas. Though Traditional Authority leaders, known as Chiefs, are highly influential in their communities for addressing harmful social and behavioral norms, they are not routinely engaged in HIV prevention planning as key actors for behavioral interventions. Consequently, programs and implementation strategies are highly fragmented and poorly tailored to the specific needs of at-risk communities in Blantyre, without public health platforms to regularly deliver prevention interventions in the community.

Blantyre's health system is not equipped to direct a dynamic and cohesive HIV prevention program. It is within this context that the Blantyre Prevention Strategy (BPS),[1] as the new HIV prevention effort has come to be known, invests in key enablers of the health system in order to strengthen local governance, improve community linkages, and bolster the success of existing partners for a more efficient, cohesive, and coordinated HIV prevention response. These enablers are: streamlined district-level governance capable of effective coordination of HIV prevention activities across public, private, and donor sectors; strengthened technical leadership; user-centered data systems that enable data users to make more informed decisions to guide the response;

---

1. The Blantyre HIV Prevention Strategy is funded by the Bill & Melinda Gates Foundation. Implementation began May 2020 and will continue through 2025. At the time of writing, the Blantyre Prevention Strategy was six months into implementation. Some activities described in this chapter are forward looking and will take place over the next five years of implementation.

routinized community and civil society engagement to understand evolving community preferences and effectively deliver services outside of health facilities; and strong multi-sectoral partnerships to utilize private sector entities for financial and technical assistance and coordinated service delivery. By investing in critical elements of the health system, the outcomes of the BPS extend beyond HIV to build local-level capacity for a rapid response — achieved by optimizing surveillance and strengthening channels for delivering prevention to those outside of routine contact with the health system. As a result, the BPS is a catalyst for transformative action that benefits other health-related conditions and health emergencies within the district.

The Blantyre Prevention Strategy

Health system investments made by the BPS are designed to optimize and institutionalize HIV primary prevention efforts along the HIV prevention cascade of targeting, demand-generation, delivery, interventions; support continued and effective use of prevention tools during periods of risk [4]; and unlock key capabilities of local actors critical for sustained program performance. The BPS embeds systems and capabilities within the district to identify with precision where new HIV infections are likely to emerge, leveraging new data on HIV risk factors and prevention indicators, along with novel data streams, and new methods for analyzing routine data. Increasing local accessibility to that data from the community-to national-level facilitates a highly localized assessment of risk for a more targeted HIV response. The BPS plans to enhance linkages across communities, civil society organizations, city and district leadership, and national decision-makers to capture local innovation and institutionalize community preferences and perceptions in program planning for improved demand-generation and delivery. Elements of the BPS strengthen district-led planning, coordination, and implementation of HIV prevention programming resulting in a strengthened health infrastructure capable of maintaining core functions and responding to other health threats and disease outbreaks.

Improving surveillance to identify risk and target services

In many settings, the detection of and response to emerging patterns of new HIV infections has been a donor-funded effort, often run outside of local systems and capacities. Some vertical surveillance systems are not mandated to report into the District Health Information System (DHIS2) and district-led surveillance efforts are often constrained by an over-burdened health workforce and incomplete reporting, restricting the District's ability to recognize trends of new infections or increased risk for HIV acquisition and respond in a rapid and coordinated manner that maximizes existing program efforts and resources.

To address this, the BPS leverages the district-based Integrated Disease Surveillance and Response (IDSR) unit. The goal of IDSR in Malawi is to improve the ability at all levels of the health system to detect and respond to diseases, conditions, or events of public health importance by providing timely and reliable data for taking action. While the IDSR guidelines include HIV and AIDS case detection, these efforts in Blantyre currently do not extend beyond the facility level and are further challenged by incomplete monthly reporting and limited ability to track new cases in real-time and target prevention services to at-risk populations [8—10]. Utilizing existing IDSR infrastructure, the BPS leverages community-based surveillance systems that have been successful for cholera, tuberculosis, and other priority diseases for HIV surveillance and rapid response teams for the immediate investigation of communities reporting either an uptick in HIV diagnoses or proxy risk factors for HIV acquisition.

Detection is improved through active and indicator-based surveillance. Routine HIV testing data collected through health facilities and communities in Blantyre is combined with novel data sources to better understand key drivers of new HIV infections according to specific sub-population and geographical locations within the district. Indicators to detect where a cluster of infection is taking place and identify proxy risk factors for HIV acquisition in a geographical area, such as an increase in teenage pregnancy, increased STI diagnosis, or low coverage of prevention interventions determine the response. While new data streams such as PEPFAR's use of recency testing [11], and program strategies such active index testing have the potential to strengthen our understanding of where new infections are emerging and facilitate responses, these systems are still being tested. As these strategies come online, having a unit at the district-level with the capacity to coordinate all surveillance efforts and direct a cohesive response will bolster success.

Active surveillance system involves the use of key personnel at the community level who are able to identify risk and make timely, necessary interventions. The implementation of Malawi's National Community Health Strategy (2015—20) has strengthened surveillance by allowing community health workers such as health surveillance assistants (HSAs) and health diagnostic assistants (HDAs) to screen for priority diseases in communities and strengthen linkage and referral to facilities [12]. HSAs and HDAs are often the first line of contact between communities and the health system and are well placed to provide critical insight into high-risk communities to inform where HIV prevention interventions should be targeted.

Through the BPS, a cadre of HSAs trained to detect proxy risk for HIV infection at the community and facility level and capture data on individuals who test negative for HIV in order to provide insights into behaviors, risks, and notable locations for future program priorities are being deployed in the district to enhance detection and response efforts. As part of their day-to-day core responsibilities, HSAs provide oversight of community health interventions in their communities including supervision, mentorship and coaching of Village Health Committees, and Community Health Action Groups. The cadre of HSAs orients these groups to HIV risk and prevention and works with district health staff to enhance community surveillance structures and identify sources within the community that may know more about the community's health status (e.g., shop owners, traditional healers, school teachers, village leaders). For example, events, risk factors, and observed attitudes, perceptions, or behaviors that could affect uptake and adherence to HIV prevention may be reported through these structures.

As risk or infection is detected, response activities that include the verification, notification, and timely deployment of prevention interventions will be initiated through the IDSR unit. Upon notice of detection —reported directly from either community surveillance informants, IISAs, other community healthcare workers, or from the IDSR unit's data review — members of a dedicated HIV rapid response team within the District Health Office travel to the community to verify and investigate the possible outbreak. Based on the results of the investigation, implement control and prevention measures are deployed. As the ability to detect new HIV cases through recency testing becomes available, voluntary assisted partner notification will be implemented per national guidelines to gather information about social contacts and understand transmission dynamics in the community to help guide additional prevention interventions at the community level. Embedding the ability to detect and respond to new or increased HIV infection within a district unit equipped with existing surveillance and response mechanisms streamlines district-level governance and builds the ability of sub-national government to direct resources and coordinate response efforts across partners.

Together, these tools and interventions allow for Blantyre's leadership to better predict and respond to patterns of transmission with greater precision. Further, the strengthened IDSR unit can be leveraged for greater preparedness for other infectious disease outbreaks — breaking patterns of transmission deep in communities before an outbreak spreads.

Improving local governance through better data use for a comprehensive response

Initial outreach directed through the IDSR unit needs to be followed by a sustained HIV prevention response to control the epidemic over the long-term. Data on diagnostics and risk must be met with data on multi-sectoral determinants of health and strong governance for comprehensive HIV prevention planning along the HIV prevention cascade. Others have demonstrated how improved understandings of HIV transmission patterns can inform efforts to address supply-side, demand-side, or adherence barriers to prevention [13].

Further, understandings of HIV transmission have evolved over time to include social and structural determinants of incidence. In Blantyre, high rates of gender-based violence, discrimination of people based on sexuality, limited access to family planning and contraceptive services, high levels of poverty and unemployment, and failure to complete secondary schooling contribute to high rates of HIV among certain subsets of the populations — namely adolescent girls and young women, female sex workers, and men who have sex with men. Population mobility and high levels of migration can also lead to increased transmission; patterns in the time and place of population migration can be predicted through data on food security, weather, and economic opportunity. These factors underscore the need for HIV prevention program managers and decision-makers to know current epidemic trends, intervention coverage, determinants of unmet needs (supply-side, demand-side, or adherence barriers), and impact to make effective and resource-efficient decisions on where, how, and to whom prevention services should be delivered.

The BPS supports District and City Councils as they make decisions and lead a cohesive response through the provision of enhanced data portals that link data from different sources. It also prepares them for automated, routine, and ad hoc investigation at all levels of the health system. To expand the capabilities for data-driven prevention decision making and enable a response to social, behavioral, and structural drivers of transmission across the wide range of government, civil society, and other partners, the BPS, working with the Ministry of Health, will integrate national data platforms and sources into an HIV prevention data pipeline.

The data pipeline enhances data use for different decision-makers by deploying user-centered data dashboards that give decision-makers a comprehensive understanding of epidemic trends (Fig. 5.2). Data from Malawi's Demographic and Health Survey, Population-based HIV Impact Assessment, and PLACE survey — which provide insights into key population size estimates, HIV "hotspot" venues and service coverage, and incidence rates disaggregated by gender and age — will be linked with health facility and community-level data required for tailoring prevention programs locally and make efficient resource allocation decisions (Table 5.1). Several emerging innovative data sources for tracking population mobility (e.g., food security, weather patterns, etc.) and the use of mobile network operator data to track individual movement, will further enhance surveillance efforts and the ability to predict outbreaks and target services. The layering of population movement data, epidemiological information, and geographic data on HIV hotspots enables more effective titration of scarce resources for HIV prevention and intensification of outreach activities during peak periods of risk.

The BPS plans to orient City Councilors on the determinants of HIV and other health issues and provide access to ward-specific data, including HIV incidence outcomes and multi-sectoral social and structural aspects and determinants, such as regulatory and bylaw enforcement to manage "hotspot" venues, insights on sex work and degree of mitigation, and crime rates with a

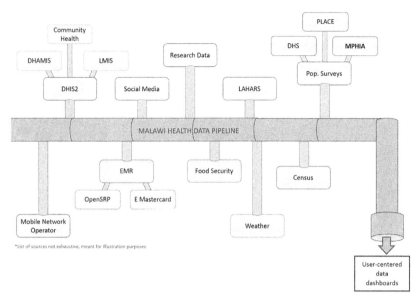

FIG. 5.2 HIV prevention data pipeline. ©*Cooper/Smith.*

focus on populations vulnerable for HIV (gender-based violence, trafficking of youth). Access to this data empowers city officials to address structural determinants of HIV transmission and contribute to an environment that promotes better health outcomes.

Traditional leaders and chiefs are essential in addressing social norms that act as a barrier to prevention seeking behavior [14]. However, strategic links these stakeholders are often missing in local health systems governance and in particular in heavily donor-driven HIV responses. This limits effective and sustainable deployment and uptake of primary prevention technologies and interventions to targeted populations outside of health facilities. The BPS addresses this gap by strengthening the engagement of Traditional Authority leaders and existing local community structures like Village Health Committees and Area Development Committees in HIV prevention planning. Supporting better access to data for these groups through the pipeline harnesses local innovation and promotes community involvement in primary prevention and response efforts.

Management of the prevention response by district and city health officials is often hampered by limited visibility of the location, mandate, activities, and services provided by community-based organizations, private medical providers, implementing partners, and corporations within a given ward. Operating without knowledge of these capacities and resources limits the government's ability to plan and leverage resources. To improve coordination of actors and activities outside of health facilities for a targeted response, a

TABLE 5.1 Illustrative list of data sources to be included in the HIV prevention data pipeline.

Data source	Description
Malawi Health Facility Registry	Comprehensive list of health facilities in the country, maintained by the Ministry of Health and Population
Integrated Logistics Management Information System	Facility commodity reporting for supply chain management
OneHealth Surveillance	IDSR reporting and surveillance data
Population based surveys	Inclusive of the Malawi Population-based HIV impact assessment, 2015–16; Violence Against Children and Young Women in Malawi, 2013; PLACE Report Malawi, 2018; Malawi Biological and Behavioral Surveillance Survey, 2013–14; Malawi Demographic and Health Survey, 2015–16
Local Authority HIV and AIDS Reporting System	HIV/AIDS non-biomedical activities implemented at district level
District Health Information System (DHIS2)	Open-source, web-based health information management system platform
Mobile network operator	Anonymized mobile network operator data can be sued as a proxy for population density and tracking population movement
Food security	Data enable prediction on population migration
Weather	Data enable prediction on population migration
Census	Population count
Social media	Information on social networks to predict at-risk populations

district-wide mapping of HIV-implementing partners, civil society and community-based organizations, and the private sector is being undertaken by BPS partners. As mapping is scaled up throughout the district, increased visibility of service providers and their activities will inform a comprehensive understanding of which services are being provided, where and to whom, and where there are gaps in the networking and referral system. The mapping data will be shared with all actors, including donor-supported clinics and programs, to enable greater coordination and ensure resources are used efficiently to maximize impact.

The coordination of actors and availability of data equips Blantyre's city and district leadership with the information needed to identify gaps in prevention delivery to targeted, at-risk populations and determine whether the unmet need is due to demand-side, supply-side, or adherence barriers. Decisions can then be made to allocate resources to an intervention or set of interventions most impactful for addressing the determinant of unmet need. These include social, behavioral, structural, and biomedical interventions [13]. Further, the BPS seeks to embed the tools and mechanisms needed for determining and delivering these interventions into local systems for institutionalized response capacity.

Greater involvement of communities and Traditional Authority leaders in HIV planning through data sharing and consistent community mobilization will be leveraged for generating program indicators aligned with community preferences. These actions, in turn, are expected to increase community buy-in and improve feedback responsiveness for program measurements. Program impact is monitored routinely and reported through the District and City Technical Working Group quarterly meetings and fed up to the national-level where program evidence can be translated into effective policy change.

Another distinguishing feature is that the BPS is informed by lessons learned from other large HIV prevention programs. For example, India's Avahan program emphasizes using real-time data to adapt programs for greater impact [15,16]. This pillar of the BPS that also supports replication of the strategy in other, similar localities.

Generating demand for HIV prevention among high-risk populations

The global focus on biomedical approaches to HIV prevention has largely been on treatment as prevention (TasP), first achieved through the prevention of mother-to-child transmission (PMTCT). The Government of Malawi demonstrated leadership, at the national-level, in pioneering "option B+" — life-long treatment for pregnant women, which has manifested in updated global clinical guidelines for clinic-based PMTCT. As of 2020, 95% of pregnant women with HIV in Malawi are on HIV treatment [1]. In recent years, TasP has shifted to treatment initiation for all people living with HIV as a key element of comprehensive HIV prevention as people with undetectable viral loads are unable to pass on the virus. With support from the Global Fund, and now PEPFAR, Malawi has achieved success with a largely clinic-based treatment response that has contributed to over 800,000 individuals accessing HIV treatment out of an estimated 1 million living with HIV [1]. The early success of the program has been attributed to a Ministry of Health-led process of evaluating clinic-based program scale-up, retention, and performance on a quarterly basis.

However, while effective biomedical interventions specifically for HIV prevention have emerged in the past decade due to large investments by global donors, scale-up has yet to match the success of the TasP approaches. Voluntary medical male circumcision is a highly cost-effective HIV prevention intervention offering a life-long reduction in HIV acquisition risk for men; modeling studies conducted in 2009−11 established that 80% coverage in 13 high-burden African countries would prevent 3.4 million HIV infections over 15 years [17,18]. Yet, uptake of VMMC has been limited with only 28% achievement of annual VMMC goals achieved in Malawi in 2018 [3]. Likewise, scale-up of oral PrEP, shown to reduce the risk of HIV acquisition by up to 99% when taken daily [19], has failed to meet global targets and national guidelines were only in early implementation in late 2020.

Increasing motivation for primary prevention intervention use, especially among marginalized populations who are at increased risk of acquisition, is critical for driving a decline in new infections [4,13,15]. Demand-side interventions are widely used to overcome behavioral and social barriers but activities are often intervention-specific, costly, and time consuming. Significant resources have gone into demand-generation activities for HIV primary prevention services and have involved private sector approaches like human-centered design (HCD) to gain insight into the attitudes, behaviors, beliefs, and motivators of community members, specifically those at increased risk of HIV acquisition. An evaluation of demand-generation activities for VMMC found understanding and responding to the precise needs and concerns of various community groups to be critical [17]. These lessons can be applied to demand-side barriers for any primary prevention intervention but approaches need to be adapted in order to be applied routinely in local planning.

The BPS adapts HCD methods for regular use in local planning and as a way to engage communities more fully. The importance of community engagement for understanding underlying drivers of HIV risk and motivators or barriers for accessing HIV prevention services is becoming evident [15,20]. In an HCD model, at-risk populations identified through intensified surveillance efforts are brought together with a capacitated civil society and community-based organization network, local service providers, and city and district health leadership to examine the unique challenges faced by that population and elicit locally acceptable solutions to issues with behavioral, biomedical or social interventions. For instance, adolescent girls (i.e., 10−19 years of age) comprise almost a third of new HIV infections in Malawi due to structural gender inequalities and multisectoral determinants such as gender-based violence, high rates of teenage pregnancy, and low levels of primary school completion [1,5]. As a result, comprehensive strategies are needed for delivering prevention interventions to this group, yet fragmentation among community-based organizations, private sector, and government-led services has constrained the ability to coordinate and integrate services for a comprehensive response. Routine elicitation of community preferences and feedback allows local prevention planners to gain a deeper understanding of the barriers faced by this vulnerable group and their attitudes and perceptions toward HIV risk so that services can be better tailored to meet their needs, thereby improving uptake and effective use of prevention products and interventions.

The insight and information elicited from communities constitute an innovative new data stream that can be systematically incorporated into novel quality improvement (QI) for prevention programs to further improve uptake and effective use of prevention interventions at the facility level in the city. QI methods have routinely been applied to HIV treatment services in Malawi but not to prevention activities. However, evidence is emerging from Southeast Asia on the effectiveness of QI methods for reducing clinic-based stigma and discrimination, a pervasive demand-side barrier in most HIV/AIDS settings that disproportionately affects young women and key populations, driving these potential high-impact users of prevention technologies further from uptake [21,22]. The BPS builds capacity within the district health management team for the application of QI approaches toward facility-based stigma reduction, PrEP, structural prevention, and VMMC delivery, and other community-identified barriers to seeking services. As the quality for prevention services improve and adapt to match user preferences, so too will the demand for the services — improving the reach and effectiveness of the program.

Public-private partnership for public health communications

Routine and accurate public health messaging is another important intervention for overcoming demand-side barriers, as it can improve risk perception and promote health seeking behavior. This too can be costly to implement and is often left to implementing partners with specific priorities that may not always be aligned with national strategies or local needs. The BPS addresses this challenge by leveraging the city's emerging commercial sector to generate routine health messaging strategies that are well-aligned with community health priorities at the local-level. Private sector marketing strategies have long-since utilized multiple channels of mass communication to influence personal decisions, yet their expertize has not been routinely captured for public health use, especially in donor-reliant health environments. Through the BPS, local corporations and businesses are brought together through Malawi's Business Coalition against AIDS and Malawi Confederation of Chambers of Commerce and Industry to develop a platform for consolidating and enhancing health communications across the district. The establishment of a health communications platform further supports routine district-level public health messaging by utilizing sustainable private sector financing sources such as corporate social responsibility programs. The communications platform can also match private sector approaches (e.g., market segmentation and advertising), with evidence-based prevention needs to improve risk perception, HIV prevention knowledge, and information on existing and novel HIV prevention products.

As private sector engagement in HIV prevention increases through routine data sharing, the sector will be challenged to intervene upon demand-side barriers in other ways, such as mitigation of HIV and health risks associated with large-scale worksites, enhanced corporate health programs and community outreach, and entrepreneurship opportunities for youth and women.

Expanding delivery channels and improving coordination to meet demand

Successful delivery of HIV prevention requires providing targeted services to large numbers of people who otherwise have infrequent contact with health facilities. Community-based organizations have played a critical role in improving adherence outcomes for HIV treatment and have been successfully utilized to deliver demand-side interventions for improving risk perception and knowledge of prevention interventions to at-risk populations who have infrequent contact with health facilities [23]. These organizations are well placed to directly support the delivery of certain biomedical prevention interventions (e.g., condoms), provide referrals to facilities for other interventions (e.g., PrEP or VMMC procedures), and address other social and structural determinants.

In the city of Blantyre, delivery among community-based organizations and linkage to facilities is challenged by the limited technical and managerial capacity and resources. The BPS recognizes the importance of coordinated community organization involvement and plans to leverage existing networks of community-based and civil society organizations in the district to draw on their expertise to systematically strengthen all HIV-related community-based organizations throughout the district. District visibility into the activities and capacity of these facilities is currently limited by incomplete and untimely reporting of organizations through the Local Authority HIV and AIDS Reporting System, which generates monthly reports of non-biomedical HIV activities from community organizations at the district level.

The BPS improves the reporting of these organizations by creating a learning environment among networked organizations to share expertise in data collection and management. Further, as reporting is eventually strengthened, the data will be integrated into the HIV prevention data pipeline for routine use. Contextually irrelevant prevention indicators exclusive of community preferences can demotivate already under-capacitated organizations to complete reporting. Establishing a data inclusive environment that addresses these challenges and routinely shares data back from facilities to community-based organizations will also enable greater linkage of community-based organizations to health facilities across the district. The linkages built between community, health site, and local government decision-makers works in both directions — serving as a mechanism for generating increased demand for services by raising awareness of risk at the community level and promoting community voice in intervention design, resource allocations, and policy planning for prevention programming.

Malawi's community health workforce is well placed to work with the capacitated and networked community-based organizations to intensify HIV prevention delivery efforts outside of facilities and improve linkage and referrals to facility-based programs relevant to at-risk population needs. Increasing recognition of the importance of community involvement for reaching HIV goals has led to the expansion of community health worker responsibilities and fragmentation among specific donor-led initiatives [24]. The BPS, aligned with Malawi's National Community Health Strategy, supports the harmonization of a community health response for accelerating HIV prevention goals. The BPS will support the District Health Office to coordinate community health efforts toward the HIV response and build a strengthened community health workforce for achieving other health goals to which community health workers are essential, such as universal health coverage and improved pandemic preparedness [25].

Greater involvement of communities in HIV prevention planning, together with the comprehensive range of health system improvements made through the BPS, will create an environment supportive of prevention seeking behavior for those at risk of acquisition, access to prevention services, and effective, and sustained use of prevention products and interventions. The involvement of traditional leaders and City Councilors in HIV prevention planning through improved data sharing will lead to evidence-informed decisions, social guidance, and policies that address social and structural barriers to health and well-being, beyond HIV.

Expanding HIV prevention delivery through private healthcare providers

The potential of private healthcare providers — such as private hospitals, provider networks, and private practitioners — is under-utilized to expand the uptake of effective prevention interventions. This is, in part, because of poor coordination with government-provided services. The last mapping of private health providers in Malawi took place in 2013 and found that approximately 40% of health services in the country are provided through the private sector, with over half of these facilities offering at least one HIV/AIDS service [26].

The BPS plans to support a process by which the services of private sector health providers can be leveraged to bolster HIV prevention and, eventually, other health services. This begins with mapping exercises to establish a baseline understanding of the gaps in their practices and services relating to HIV prevention. The BPS capacitates private sector providers by certifying them through the district as a provider of advanced HIV prevention, and by linking them to ongoing training on novel HIV prevention methods and interventions. To further improve district coordination, private facilities are linked to monitoring and evaluation systems to enable the tracking of HIV prevention intervention (e.g., PrEP, VMMC) coverage rates within private healthcare networks. Importantly, given the rapid rate at which new private providers are entering into the marketplace, the program will be offered to new registrants, which ensures a sustained pool of private providers engaging in high-impact HIV prevention service delivery.

Conclusion

The city of Blantyre is not unique among major metropolitan areas — particularly in Sub-Saharan Africa — in facing a significant HIV epidemic. Thus far, the city has not experienced the significant reductions in HIV infections enjoyed elsewhere, even as substantial progress has been made in HIV treatment. Boosting the local and national governments' systems — while simultaneously investing in the HIV prevention cascade capabilities of targeting, demand-generation, delivery, and sustained use — are essential for addressing the persistent reservoirs of infection and sustaining control of the epidemic in the long-term.

The BPS is designed to support the achievement of the goals outlined in Malawi's National Prevention Strategy (2018–20) and to sustain progress by institutionalizing HIV prevention systems and interventions into local structures, and by improving the coordination and performance of all HIV actors for greater impact (Fig. 5.3). Ultimately, strengthened local systems and sustained district-level capacity for coordination, resource leveraging, and planning will yield benefits beyond HIV — including broader pandemic preparedness efforts and universal health care agendas.

Elements of the BPS present multiple areas of convergence with a broader health response and greater pandemic preparedness. Historically, the HIV/AIDS response has considerably strengthened health systems in low- and middle-income countries through increased resources, improved infrastructure, and a strengthened health workforce. The BPS's focus on strengthening systems elements and addressing key gaps in governance, technical leadership, data systems and use, civil society and community engagement, and multisectoral partnerships will bolster the performance of all health programs in Blantyre. Embedding key functions within local systems and increasing multisectoral coordination, from the outset, also bolsters district-level capacity for managing a major epidemic response. This stronger, district-led coordination of programming provides protects against the disruption to routine services in times of crisis, when strong international and domestic pressures can result in trade-offs in under-capacitated health systems. In doing so, the BPS provides local leadership the capacity and capability to set evidence-based response priorities and mobilize actors for immediate response to health emergencies.

Investments made through the BPS will increase district-level capacity for identifying and reaching at-risk populations through local systems by improving access to relevant data, strengthening links to community structures, and

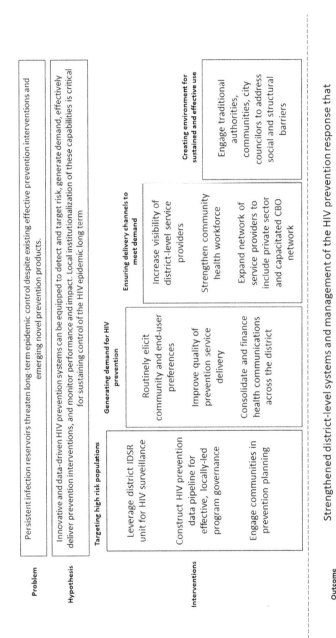

Problem

Persistent infection reservoirs threaten long-term epidemic control despite existing effective prevention interventions and emerging novel prevention products.

Hypothesis

Innovative and data-driven HIV prevention systems can be equipped to detect and target risk, generate demand, effectively deliver prevention interventions, and monitor performance and impact. Local institutionalization of these capabilities is critical for sustaining control of the HIV epidemic long term

Interventions

Targeting high risk populations

Leverage district IDSR unit for HIV surveillance

Construct HIV prevention data pipeline for effective, locally-led program governance

Engage communities in prevention planning

Generating demand for HIV prevention

Routinely elicit community and end-user preferences

Improve quality of prevention service delivery

Consolidate and finance health communications across the district

Ensuring delivery channels to meet demand

Increase visibility of district-level service providers

Strengthen community health workforce

Expand network of service providers to include private sector and capacitated CBO network

Creating environment for sustained and effective use

Engage traditional authorities, communities, city councilors to address social and structural barriers

Outcome

Strengthened district-level systems and management of the HIV prevention response that maximizes and sustains investments towards programmatic goals

FIG. 5.3 Blantyre's HIV prevention strategy.

increasing multisectoral coordination for comprehensive response planning. Local insights and pathways for community-driven approaches are essential for developing an optimized system and response for any disease. Pathways for incorporating community-driven approaches in government and partner planning will be realized in the City, providing a continuous process for generating local solutions to unique community challenges for disease writ large.

The BPS was developed prior to the COVID-19 pandemic, and like other public health approaches, has had to reckon with the impact on its planned activities. Mitigation of COVID-19 infection, morbidity, and mortality are critically important in the midst of the crisis as is the protection of the tremendous gains Malawi has made in HIV treatment and viral suppression toward the achievement of the 90-90-90 goals. Ensuring HIV prevention tools and services are available to those who need them is essential to avoiding an increase in new HIV infections. Existing elements of the BPS are well-placed to support the COVID-19 response in Blantyre, such as the strengthening of the district's IDSR unit and community surveillance structures, and efforts to increase the ability of local decision-makers to analyze novel data for population mobility tracking and target response interventions. Additionally, the communications channel established through the public-private partnership will be utilized for rapid and accurate public health messaging. Perhaps most importantly, enabling locally relevant solutions to the delivery of interventions to populations outside of routine contact with health facilities supports the rapid deployment and uptake of novel biomedical prevention products as they come to market.

The BPS will develop a replicable model that optimizes and sustains HIV prevention performance and benefits other district-led health programs. Further, the efforts to monitor the performance of the BPS through real-time data allows for successful elements of the BPS to be scaled rapidly in other priority locations throughout the country and lessons will be applied to geographies with similar epidemics and contexts throughout the region. With implementation taking place during a public health emergency, the BPS has a unique opportunity to document what effect COVID-19 has on HIV prevention services, how to mitigate negative impact, and how to dual-purpose activities to support the HIV and COVID-19 responses. Ultimately, the BPS codifies practices that can support enhanced health system resilience in the event of future health emergencies.

References

[1] UNAIDS. Country factsheet Malawi. 2019. aidsinfo.unaids.org. [Accessed 1 October 2020].
[2] National AIDS Commission, Malawi. Revised national HIV prevention strategy 2018-2020. Lilongwe: National AIDS Commission; 2018.
[3] UNAIDS. HIV prevention coalition Malawi country scorecard, 2019. 2019. hivprevention coalition.unaids.org/global-dashboard-and-country-scorecards. [Accessed 3 April 2020].

[4] Schaefer R, Gregson S, Fearon E, Bernadette HB, Hallett TB, Hargreaves JR. HIV prevention cascades: a unifying framework to replicate the successes of treatment cascades. Lancet HIV 2019;6:e60−66.

[5] Ministry of health of Malawi, United States centers for disease control and prevention (CDC), ICAP at Columbia University. Malawi population-based HIV impact assessment (MPHIA) 2015-16: final report. Lilongwe: Ministry of Health; 2018.

[6] National Statistical Office of Malawi. Malawi population and housing census report − 2018. Lilongwe: National Statistical Office; 2019.

[7] President's Emergency Plan for AIDS Relief (PEPFAR). Malawi country operational plan, COP 2019, strategic direction summary. Lilongwe: PEPFAR; 2019.

[8] Wu TSJ, Kagoli M, Kaasbøll JJ, Bjune GA. Integrated disease surveillance and response (IDSR) in Malawi: implementation gaps and challenges for timely alert. PLoS One 2018;13:e0200858.

[9] World Health Organization. Joint external evaluation of IHR core capacities of the Republic of Malawi. Geneva: WHO; 2019.

[10] World Health Organization. Integrated disease surveillance and response technical guidelines, booklet one: introduction section. Brazzaville: WHO Regional Office for Africa; 2019.

[11] Kim AA, Behel S, Northbrook S, Parekh BS. Tracking with recency assays to control the epidemic: real-time HIV surveillance and public health response. AIDS 2019;33:1527−9.

[12] Ministry of Health of Malawi. National community health strategy 2017−2022. Lilongwe: Ministry of Health; 2017.

[13] Hargreaves J, Delany-Moretlwe S, Hallett TB, Johnson S, Kapiga S, Bhattacharjee P, et al. The HIV prevention cascade: integrating theories of epidemiological, behavioural, and social science into programme design and monitoring. Lancet HIV 2016;3:e284−5.

[14] Downs JA, Mwakisole AH, Chandika AB, Lugoba S, Kassim R, Laizer E, et al. Educating religious leaders to promote uptake of male circumcision in Tanzania: a cluster randomised trial. Lancet 2017;389:1124−32.

[15] Sgaier SK, Ramakrishnan A, Wadhwani A, Bhalla A, Menon H, Baer J, et al. Achieving scale rapidly in public health: applying business management principles to scale-up an HIV prevention program in India. Healthcare 2018;6:210−7.

[16] Tran NT, Bennett SC, Bishnu R, Singh S. Analyzing the sources and nature of influence: how the Avahan program used evidence to influence HIV/AIDS prevention policy in India. Implement Sci 2013;8:44.

[17] Sgaier SK, Baer J, Rutz DC, Njeuhmeli E, Seifert-Ahanda K, Basinga P, et al. Toward a systematic approach to generating demand for voluntary medical male circumcision: insights and results from field studies. Glob Health Sci Pract 2015;3:209−29.

[18] Njeuhmeli E, Forsythe S, Reed J, Opuni M, Bollinger L, Heard N, et al. Voluntary Medical Male Circumcision: modeling the impact and cost of expanding male circumcision for HIV prevention in eastern and southern Africa. PLoS Med 2011;8:e1001132.

[19] United States Centers for Disease Control and Prevention. HIV, pre-exposure prophylaxis (PrEP). 2019. www.cdc.gov/hiv/risk/prep/index.html. [Accessed 10 February 2020].

[20] Joint United Nations Programme on HIV/AIDS (UNAIDS). Global AIDS update 2019 − communities at the centre. Geneva: UNAIDS; 2019. Available from: https://www.unaids.org/sites/default/files/media_asset/2019-global-AIDS-update_en.pdf.

[21] Kapanda L, Jumbe V, Izugbara C, Muula AS. Healthcare providers' attitudes towards care for men who have sex with men (MSM) in Malawi. BMC Health Serv Res 2019;19:316.

[22] Ikeda DJ, Nyblade L, Srithanaviboonchai K, Agins BD. A quality improvement approach to the reduction of HIV-related stigma and discrimination in healthcare settings. BMJ Glob Health 2019;4:e001587.

[23] Wouters E, Van Damme W, van Rensburg D, Masquillier C, Meulemans H. Impact of community-based support services on antiretroviral treatment programme delivery and outcomes in resource-limited countries: a synthetic review. BMC Health Serv Res 2012;12:194.

[24] De Neve JW, Garrison-Desany H, Andrews KG, Sharara N, Boudreaux C, Gill R, et al. Harmonization of community health worker programs for HIV: a four-country qualitative study in Southern Africa. PLoS Med 2017;14:e1002374.

[25] Tulenko K, Møgedal S, Afzal MM, Frymus D, Oshin A, Pate M, et al. Community health workers for universal health-care coverage: from fragmentation to synergy. Bull World Health Organ 2013;91:847—52.

[26] United States Agency for International Development (USAID). Malawi private health sector mapping report. Washington: USAID; 2013.

Chapter 6

Building a robust interface between public health authorities and medical institutions in a densely populated city: state-of-the-art integrated pandemic and emerging disease preparedness in the Greater Tokyo Area in Japan

Takako Misaki[1], Tomoya Saito[2], Nobuhiko Okabe[3, 1]

[1]*Division of Planning and Management, Kawasaki City Institute for Public Health, Kawasaki, Kanagawa, Japan;* [2]*Center for Emergency Preparedness and Response, National Institute of Infectious Diseases, Shinjuku-ku, Tokyo, Japan;* [3]*Director General, Kawasaki City Institute for Public Health, Kawasaki, Kanagawa, Japan*

Japan is an island country located to the east of the Eurasian Continent. As many as 126 million people live on the land of only 378,000 km^2, and more than 30 million foreigners visited Japan every year before the COVID-19 outbreak. Haneda International Airport is located in Tokyo, the capital of Japan, and 39,000 flights arrive at or depart from the airport annually.

Kawasaki City is a large city in the Metropolitan area of Japan, and is located between Tokyo with a population of 13.85 million to the west, and Yokohama with a population of 3.74 million to the east (Fig. 6.1). Kawasaki

---

[1] This article was developed with support from the Health and Welfare Science Research Fund by the Ministry of Health, Labour and Welfare. This article contains the author's personal analysis and views and does not represent the official position of the organization.

Inoculating Cities. https://doi.org/10.1016/B978-0-12-820204-3.00006-1

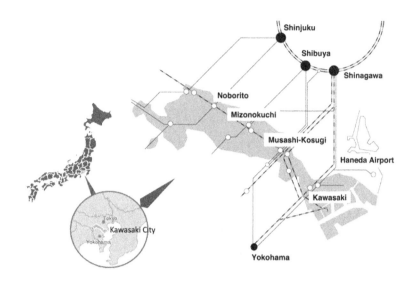

Kawasaki City

FIG. 6.1   The city of Kawasaki, Kanagawa Prefecture, Japan.

City consists of seven wards, with a total population of 1.53 million — a large proportion of which travels to and from central Tokyo on a daily basis. It is also a government-designated city with its own authority to control infectious diseases through local public health centers.

The city is constantly exposed to the threat of an emerging infectious disease outbreak. It has an international seaport and is close to Haneda International Airport. The proportion of foreign residents exceeds 6%, a relatively high percentage in Japan. From a historical perspective, Kawasaki City reported the first case of influenza in the Metropolitan area during the 2009 influenza pandemic and also reported the first case of Zika virus infection after the Ministry of Health, Labour and Welfare of Japan (MHLW) designated it as a notifiable disease in 2016.

Kawasaki City has taken several steps to prepare itself to respond to an infectious disease crisis. In Japan, a robust indicator-based surveillance structure has been established based on the Law on the Prevention of Infectious Diseases and Medical Care for Patients with Infectious Diseases (hereafter the Law) [1]. Physicians and medical institutions are obliged to report specified infectious diseases by the Law. However, there have been challenges in detecting emerging infectious diseases that are not classified as notifiable diseases. Reinforcing "event-based surveillance" by establishing systems for early reporting from medical institutions and risk assessment is crucial for addressing these challenges. To make this type of system work — to detect an event and rapidly respond to the threat — an established mechanism for collaboration between public health authorities and medical institutions is a key. To this end, Kawasaki City has established real-time and interactive information sharing systems with local medical institutions. It has also nourished a trusted relationship between the local health department and medical institutions through joint exercises. The city has also established a unique human resource development program, the Field Epidemiology Training Program-Kawasaki to develop core human resources and bolster the municipal staff's risk assessment capabilities. No-notice mystery patient exercises have validated the relationships between medical institutions and the local health department for early detection and response. Finally, the Act on Special Measures for Pandemic Influenza and New Infectious Diseases Preparedness and Response (the Special Measures Act) [2], enforced 2013, legislated social distancing measures and mass vaccination programs. As a means of implementing this Act, Kawasaki City piloted a mass vaccination program for pandemic influenza and proposed a mass vaccination model for other urban areas.

Kawasaki City has established robust and advanced systems to respond to infectious disease crises through these efforts and presented a state-of-the-art model for urban infectious disease control in Japan. This chapter further details these initiatives of Kawasaki City.

Surveillance systems in Japan and the Kawasaki City Infectious Disease Surveillance System

The authority of infectious disease control at the local-level in Japan is under each prefecture. Large cities, such as government-designated cities (major cities), core cities (mid-level cities), and some special wards with their local public health centers have responsibilities relating to disease control. Kawasaki City is one of the 20 government-designated cities. It has its own local public health center, with branches in each of the city's seven wards, that undertakes measures in response to infectious diseases. Kawasaki City also has its own public health institute (the Kawasaki City Institute for Public Health; KCIPH), which conducts diagnostic testing necessary for disease control under the Law and houses the Kawasaki Infectious Diseases Surveillance Center (KIDSC).

The Implementation Manual for the National Epidemiological Surveillance of Infectious Diseases Program urges public health institutes to establish local infectious disease surveillance centers (IDSC) to collect and analyze information on notifiable diseases as outlined by the Law. As of January 1, 2020, there were 83 public health institutes in Japan that participate in the National Epidemiological Surveillance of Infectious Diseases (NESID) program in collaboration with the central IDSC in the National Institute of Infectious Diseases (NIID).

Under the Law, medical institutions collect clinical samples from patients that are then tested in laboratories. Microbiological tests that are required by the notification criteria but are not available at private labs are performed at the local public health institutes. The local IDSC collects and analyzes case reports submitted to authorities from physicians through local public health centers and publishes reports on a weekly and monthly basis. The public health institutes also report the pathogen information and test results to NIID, which, in turn, analyzes and publishes nationwide reports. For example, the Infectious Diseases Weekly Reports publishes the national data on the number of flu patients per sentinel site, endemic viral types and strains on influenza based on reports from the local public health institutes. These data are further reported to the World Health Organization and included in global epidemiological surveillance reports and used as an essential reference for selecting flu vaccine strains and monitoring the emergence of new influenza viruses.

Systems for information sharing between public health authorities and medical institutions in Kawasaki - lessons from pandemic influenza 2009

In April 2009, the US and Mexico reported the emergence of pandemic influenza A(H1N1)pdm09. The disease spread to all parts of the world in a short time and on April 28, 2009 H1N1pdm became a notifiable disease. Japan identified the first in-country patient with no travel history on May 16 and the MHLW issued a notice urging local governments to strengthen surveillance and epidemiological investigations. Physicians were encouraged to report clusters and severe cases of any type of influenza. At that time, the daily number of patients with type A influenza was collected by fax from medical institutions. The Kawasaki Medical Association, an association of physicians, manually aggregated the report, which proved to be a labor-intensive process. Based on this experience, the Kawasaki Medical Association requested that the city establish an online reporting system for emerging diseases. In response, Kawasaki City developed a web-based surveillance system to make the reporting process more efficient and to enhance communication to prepare for emerging and re-emerging infectious diseases such as pandemic influenza.

The Kawasaki City Infectious Disease Surveillance System (KIDSS)

The city launched the Kawasaki City Infectious Disease Surveillance System (KIDSS) in April 2014. The KIDSS is a web-based system and open to the public (Fig. 6.2) [3]. Registered medical institutions have secured access to disease reporting and detailed information sharing. As of December 2019, nearly 70% (695/1010) of the medical institutions in the city have registered in the KIDSS.

FIG. 6.2   The Kawasaki City Infectious Disease Surveillance System (KIDSS).

The KIDSS represents a one-stop portal providing seven essential services. These include publishing data from the NESID program, real-time surveillance data, a community bulletin board, a reference library, factsheets and case definitions for notifiable diseases, syndromic surveillance data for schools and nurseries, and messaging functions. Among these, the data from NESID program, real-time surveillance data, the factsheets and case definitions for notifiable diseases, and the syndromic surveillance data for schools and nurseries can be accessed by the general public. English pages have also been created to enable access by international users.

Publication of data from the NESID program

The current epidemiological profiles of infectious diseases in the city are published in tables, graphs, and maps based on the NESID program data [3]. The general public has had access to the data since 1999 when surveillance activities began under the Law. The KIDSC updates the information every Wednesday and the system has an interactive interface that enables users to customize the views (e.g., diseases, periods, districts, etc.). Data are also available to be downloaded as CSV files for further use.

Real-time surveillance

Real-time surveillance aggregates daily information from medical institutions (e.g., the number of patients) and publishes it in tables, graphs, and maps on a real-time basis [3]. Presently, this system focuses on both type A and type B influenza. Seasonal influenza is a disease notifiable only at sentinel sites under the NESID program. Still, medical institutions in Kawasaki City contribute to this surveillance, not only for monitoring cases during the flu season, but also for maintaining this system in case it is activated for a future emergency. When a new disease that has yet to be designated as notifiable emerges, the KIDSS can establish a new case definition and implement real-time surveillance in registered medical institutions in the city. The system is also capable of conducting surveillance by ward or by the clinical department. This flexibility, therefore, has also better prepared Kawasaki City to respond to threats that are unexpected or currently unknown.

Bulletin board

Staff at medical institutions who are engaged in clinical practice are the first-line defense of detecting unusual events. The KIDSS installed the bulletin board on secured pages for sharing epidemiological updates between medical staff to facilitate the early detection of and subsequent intervention for infectious disease outbreaks. Unless it is a notifiable disease, this information may not always be reported to the local public health centers making it difficult for a single medical institution to suspect an emerging outbreak in the area if patients are disbursed among separate medical institutions. Accordingly, the KIDSS bulletin board helps clinical staff at medical institutions and the public health authorities to overcome such difficulties. Presently, only registered medical institutions and related personnel have access to posting and browsing the bulletin board. It is also possible to open specific bulletin boards dedicated to specific wards or clinical departments and the bulletin board has a push notification system for sharing new posts with specific registered medical institutions.

Reference library

The reference library provides briefs and manuals issued by both the national and local governments, Q&As, reference materials from the Health Science Council, clinical guidelines from professional societies, and other relevant materials. The contents of the library are categorized, and users from registered medical institutions are able to search the library using keywords. The KIDSS introduced this function to deliver the latest government notices directly to physicians engaged in clinical practice as a means of incorporating this important knowledge into their medical practice.

Factsheets/case definitions

The KIDSS provides notifiable disease factsheets and case definitions, as well as the criteria and notification format for infectious diseases as specified in the Law. It also contains factsheets for the general public [3]. At present, more than 110 diseases are designated as notifiable diseases in Japan, including some that are rarely (e.g., Ebola hemorrhagic fever) or never (e.g., Smallpox) observed in the country. Links are provided to the disease information at NIID's website to support and facilitate physicians' diagnosis and reporting of patients.

Syndromic surveillance in schools and nurseries

The KIDSS also provides syndromic surveillance in schools and nurseries. It publishes the number of absent children categorized by symptom and by disease, as well as the status of school and class closures at nurseries, kindergartens, and preschools in the city. Like much of the other surveillance data, these are published in a variety of formats including tables, graphs, and maps [3]. The Japanese Society of School Health operates this system where schools and nurseries record and enter absenteeism due to infectious diseases. In Kawasaki City, public nurseries began participating in this system in August 2014 and private nurseries shortly after in October 2014. Since in November 2014, the numbers of absent children in nurseries categorized by disease or by symptom entered in the system have been aggregated and published in the KIDSS. These data are automatically updated three times per day. This surveillance made the epidemiological picture of infectious diseases among local infants visible on a real-time basis, though it was limited to nurseries and preschools. In attempts to increase the number of participating facilities and improve surveillance quality the KIDSC provides startup and follow-up training, for participating nurseries. While information by symptom is provided only at the secured pages, the number of cases by diseases and class closure information is publicly available. This system was also promoted by the Ministry of Education, Culture, Sports, and Sciences to conduct syndromic surveillance for COVID-19 in schools in Japan.

Messaging

Infectious disease information that requires rapid dissemination and immediate notification can be shared with all registered local medical institutions using the messaging service in the KIDSS. Even medical institutions without internet access can receive information by fax, thereby enabling rapid and uniform information provided to all registered medical institutions. For example, when there is an accumulation of patients with diseases of unknown origin, the city can use this messaging service to quickly raise awareness among local stakeholders.

Kawasaki City created the KIDSS as a network for connecting medical institutions with the public health authorities and the system acts as a one-stop portal in the consultation rooms of registered medical institutions – enabling clinical staff to quickly review information on diseases on both endemic and more exotic diseases. Although the KIDSS was designed to be an information-sharing system for infectious diseases in local areas, it also represents an expandable platform that may be used for the early detection of other types of public health crises (e.g., natural disasters, terrorist attacks, etc.).

Field Epidemiology Training Program-Kawasaki (FETP-K): a human resource development program in Kawasaki City

Human resource development is a key capacity in local disease control; however, it's not easy for a local government to have its own program for developing specialized skills such as disease control. A Field Epidemiology Training Program (FETP) is a human resource development program of field epidemiologists who detect and respond to an outbreak. FETP is modeled on the two-year Epidemic Intelligence Service that was created by the US Centers for Disease Control and Prevention in 1951. The first FETP course was established in Canada in 1975 and has been subsequently followed by over 60 programs across the world. FETP-Japan (FETP-J) was established as a two-year field training course at NIID in 1999 to build a nationwide network of field epidemiologists. As of January 2020, 77 individuals have completed FETP-J. Twenty percent of these participants are from local governments and were expected to be engaged in infectious disease control at the local-level following the program. However, because of routine job rotations in their workplaces every few years, these staff members are usually unable to continue to develop their specialized careers in communicable disease control.

Although Kawasaki City has not sent its staff to full FETP-J, the city has recognized the importance of such a program. Fortunately, the former director of the IDSC of NIID, who was engaged in the establishment of FETP-J, was inaugurated as the Director-General of the KICPH in 2012 and one FETP-J alumnus started to work in the Institute in 2013. With these supervisors, the city established its own FETP program called FETP-Kawasaki (FETP-K) in 2014. Two or three members from the staff working at the Kawasaki Public Health Center and the KCIPH are selected every year to be trained. Importantly, these trainees include not only physicians, but also veterinarians, pharmacists, medical technologists, and other healthcare professionals. By January 2020, FETP-K had enrolled and trained fifteen staff members to improve local-level preparedness.

As part of this, FETP-K trainees join a one-month introductory training course offered by FETP-J, and then participate in a local working group as well as the FETP-K meeting at KCIPH once a month. The working group shares and discusses the local epidemic situation, which is based on laboratory and epidemiological data. The FETP-K meeting discusses the broader epidemiological situation beyond the city and occasionally conducts risk assessments and reviews of public health measures. Furthermore, FETP-K holds epidemiological seminars four times a year. These seminars include visits from supervisors and trainees to local offices in Kawasaki City to provide on-site lectures for staff in charge, with special attention given to ongoing public health events and risk assessments. The FETP-K training also requires the trainees to investigate a specific topic during their two-year training and to give a presentation at an academic conference at least once. This experience provides trainees with an important opportunity to have their work reviewed and improve their presentation skills, which may be useful for risk communication. Meanwhile, the city also provides FETP-J trainees on-the-job training opportunities with the local government and FETP-J trainees are accepted in Kawasaki City as a trainee once a week for 3 months. These types of human resource exchanges are crucial for strengthening networks in disease control between local governments and the NIID (Fig. 6.3).

Implementation of the Act on Special Measures for Pandemic Influenza and New Infectious Diseases Preparedness and Response

Although Japan had developed a national action plan, during the 2009 influenza pandemic, many challenges were identified in the public health response in Japan [4]. The lessons from this pandemic ultimately led to the enactment of the Special Measures Act, which took effect in April of 2013. The Special Measures Act provides the legal basis for planning and implementing pandemic influenza control measures, especially those relating to social distancing measures and securing necessary medical resources. In addition, the Act encourages governments to conduct training and exercises; accordingly, Kawasaki City has addressed pandemic preparedness through conducting various simulations and a mass vaccination field exercise.

FIG. 6.3 The framework of collaboration between the national FETP-J and local FETP-K program.

Simulated table-top exercise for engagement of stakeholders

A successful pandemic response must not be limited to the public health and medical sectors, but should also engage other stakeholders. Peacetime exercises and workshops where stakeholders share their roles and activities and build a network are indispensable for enhancing the multidisciplinary collaborations that are necessary for effective action during a public health emergency. Kawasaki City has hosted an annual workshop on pandemic influenza preparedness — inviting a moderator who previously contributed to the development of the National Action Plan for Pandemic Influenza and developed an exercise tool. Participants in the workshop included staff from the public health center, municipal crisis management staff, physicians and nurses engaged in infection control at medical institutions, and staff at fire departments. The workshop included a lecture that provided an overview of pandemic influenza preparedness, concepts relating to the Special Measures Act and the National Action Plan, and a simulated table-top exercise (TTX). The participants formed small groups of six to seven, and following introductions, participants were presented a scenario that transitioned between various pandemic phases. Throughout the exercise, participants roleplayed a municipal response and discussed questions raised (Table 6.1). This experience and the subsequent discussion allowed participants to share a common

TABLE 6.1 Phases and scenarios included in the table-top exercise.

Scene 1: Alert Phase 1 (Emergence of new virus overseas)	
TTX1 Ice breaking	What are your concerns about infectious disease risk management?
TTX2 Emergency declaration	What should the local government response headquarters do?
	What are the risk communication priorities for citizens?
Scene 2: Alert Phase 2 (In-country outbreak)	
TTX3 Response to first in-country patient	Should relevant schools and businesses close?
TTX4 Hospital surge capacity	How should new patient clusters be managed if designated wards are occupied?
Scene 3: Pandemic Phase	
TTX5 Social distancing measures	Should mass gathering events be canceled?
TTX6 Maintaining regional medical capacity	What measures for reducing demand and increasing supply of medical resources can be taken?

understanding of what operations could look like under the Special Measures Act and the National Action Plan. It also provided participants an opportunity to better understand their roles, the roles of others, and associated challenges.

Patient management exercises

Kawasaki City has also conducted field exercises on mass patient management. The National Action Plan specifies that during the early phases of a pandemic, when the patients are mostly limited to those who had contact with known cases or returnees from an endemic area, those at higher risk of infection should consult a call center before visiting an outpatient unit. The designated outpatient units' role is then to centralize patients to facilities with testing capacities and adequately trained for infection prevention and control (IPC) measures, thereby minimizing nosocomial infections and transmission within healthcare facilities.

Kawasaki City has proactively requested eleven medical institutions to establish these designated outpatient units prior to an emergency. In collaboration with these medical institutions, for the past two years, the city has conducted field exercises for opening call centers and tested the flow of patients through designated outpatient units. The exercises worked to verify a number of important considerations including the flow of patients from consultations at call centers to referred designated outpatient units, from a patients' home to an outpatient unit, and from the entrance to the institution to a consultation room; the assignment of physicians; testing procedures; the transport of clinical specimens; donning and doffing personal protection equipment (PPE); the preparation of a waiting room for testing results; explanations of test results to the patients and their family; and the transport of patients to a designated hospital. These exercises revealed the importance of preparation in advance of emergencies, as well as differences in the problems experienced by various hospitals due to different structures and staff assignments. Ultimately, these exercises allowed for these consultation centers and outpatient units to operate efficiently during the early response efforts to the COVID-19 pandemic.

Exercise for social distancing measures: IPC at mass gathering events in large facilities

During a pandemic or under the threat of a pandemic in Japan, under the Special Measures Act, the Prefectural Governor can restrict the use of large facilities or ask such facilities to implement IPC to prevent the spread of disease by reducing human-to-human contact opportunities. In 2018, using a local horse racetrack company as an example, Kawasaki City conducted an exercise on IPC measures such as entry-screenings for fever, hand hygiene, and disinfection at large facilities (Fig. 6.4). The exercise included many

FIG. 6.4 Kawasaki City conducted a joint exercise on the IPC measures at a local horse-race track venue (A) such as entry-screening of illness, hand hygiene, and disinfection (B) in 2018.

stakeholders including the horse racetrack company, the local public health center and Public Health Institute, the Kawasaki City General Affairs and Planning Bureau responsible for crisis management, the Kawasaki City Economy and Labor Bureau, the Security Division of the Kanagawa Prefectural Police, the Office for Pandemic Influenza and New Infectious Diseases Preparedness and Response of the Cabinet Secretariat, the Tuberculosis and Infectious Diseases Control Division of the MHLW. This exercise revealed that although many security guards at the racetrack were ex-police officers and experienced in crowd control procedures, they experienced difficulties in responding to complaints of visitors at the entrance gates, suggesting that substantial chaos could be expected during actual operations. While the restriction of the use of facilities would be feasible during an outbreak, the exercise underscored the need for advanced public relations, better crowd control, and repeated exercises.

Pandemic mass vaccination operation model in urban areas

The Special Measures Act and the National Action Plan outline procedures for vaccination campaigns for novel influenza and Kawasaki City has worked to test and validate these procedures for mass vaccinations in urban environments through field exercises.

There are many challenges in operationalizing vaccination campaigns during a pandemic in urban areas because of their large, dense populations and the abbreviated time frame. City authorities examined various types of models for operationalizing vaccination campaigns including individual vaccinations in medical institutes, group vaccinations at facilities such as schools and local communities, and vaccinations at regional medical institutions. Based on the results of the exercise, the city endorsed a hybrid model whereby those with underlying diseases, pregnant women, and preschool infants receive individual vaccinations in medical institutes; inpatients, residents in welfare facilities, school children, and other groups receive group vaccination at facilities; and all other residents undergo group vaccinations at appropriate or ad hoc venues in their local communities. The vaccination schedule strove to complete vaccinations for all residents within approximately four months from the start of vaccination. Preliminary analyses of the hybrid model exposed unique difficulties in the implementation of group vaccination at the ad hoc vaccination sites in local communities. In response to this, the city conducted an additional simulated pilot exercise for group vaccination at an elementary school and tested the city's capacity for vaccinating more than one million residents (Fig. 6.5). Considering the human resources required for preparations and operations at the ad hoc site, and the estimated shots that could be provided, the exercise demonstrated that the proposed concept was not feasible in large cities like Kawasaki City. Based on these results, the MHLW is reexamining the national vaccination strategy to make it more efficient.

No-notice mystery patient exercises

Municipalities and medical institutions have prepared many manuals and guidelines and developed training and initiatives for responding to emerging health threats. While the hope is that these processes and capacities will never need to be used, it is imperative that they be tested through exercises to better prepare for infectious disease outbreaks. Most exercises are conducted under prespecified scenarios that are provided in advance to participants. This, despite the reality that public health emergencies are almost always unique and unexpected — starting with an undiagnosed patient, with some illness, who

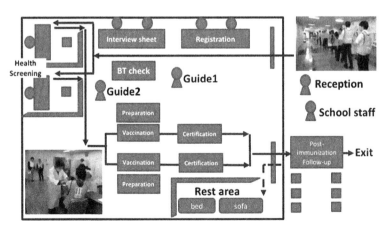

FIG. 6.5 A flow of mass vaccination operation exercise at a school in Kawasaki City, 2017. Simulated pilot exercise for "group vaccination" operation was tested at elementary school in Kawasaki City, 2017. Vaccines were registered, screened for health conditions, vaccinated and given a certificate. *BT*, body temperature.

only later will be recognized as the index case of the emergency. Indeed, while it is more complicated to plan exercises simulating an unexpected situation, in which the scenario may change over time and be influenced by the response measures taken, such exercises are much more practical.

In 2016, a "no-notice mystery patient exercise" was planned and conducted in Kawasaki City to test the hospital's capability to detect and respond to a suspected case of highly contagious disease (Table 6.2). This exercise, the first conducted in the city, relied on collaborations among several related institutions, including public health centers, hospitals, and fire departments. No other similar exercises have been conducted in Japan. The scenario was not disclosed to exercise participants except for the mock patients. Relevant medical staff were notified via telephone that an exercise would begin shortly before the start of the exercise (i.e., 15 min) or by receptionists who were informed by the mock patients.

Before the exercise, detailed profiles of simulated patients were created that included a variety of considerations such as name, age, date of birth, place of living, family composition, occupation and location of work, commute preferences, underlying diseases, clinical course (i.e., changes in vital signs and clinical symptoms every 10 min) and vaccination history. Mock patients responded to questions from hospital staff about the scenario, by providing them with simulated travel histories, symptoms, test results, and other information. The hospital responded based only on the information collected from the mock patients. All of the responses in the hospital were recorded and the maximum duration of the exercise was set at 5 h, regardless of the process or patient outcome.

Since the first no-notice exercise conducted in 2016, three others have been conducted at different medical institutions and public health centers (Fig. 6.6). The different medical institutions that have been selected for the exercise have been either university hospitals designated as core hospitals in the respective

TABLE 6.2 No-notice mystery patient exercises conducted in Kawasaki City, 2016—18.

Year	Scenario	Differential diagnosis
2016	A patient with fever and respiratory illness with a travel history to Middle East and Asia	Middle Eastern respiratory syndrome; novel influenza
2017	A patient with fever, eruption and respiratory illness with a travel history to Western Europe and Middle East	Measles; Middle Eastern respiratory syndrome
2018	A cluster of patients with a flu-like illness	Anthrax (bioterrorism)

FIG. 6.6 No-notice mystery patient exercises were conducted in Kawasaki City from 2016 to 2018 in 3 hospitals under different scenarios. Mock patients with a respiratory illness who visited the outpatient unit were taken to a temporary isolation unit at a parking lot (A), and a makeshift isolation unit in an emergency outpatient unit (B). When a cluster of patients with an influenza-like-illness visited an outpatient unit in 2018, mock patients were triaged at a waiting room (C).

secondary medical care blocs of the city, or medical institutions closely engaged with particular communities. This decision was justified based on the understanding that these hospitals are more likely to be visited patients with infectious diseases caused by unknown pathogens.

These no-notice exercises examined standard procedures for managing patients in hospitals who are at-risk for infection with an unknown pathogen. They also tested the institutes' capacity to report diseases requiring the immediate notification of public health centers. As detailed earlier, exercises are usually restricted by following a predetermined scenario and planned actions; however, in no-notice mystery patient exercises, the only available information is delivered by the mock patients. Accordingly, the simulated response is more dynamic and is dependent on the risk assessment based on the patients' history and physical examination. Although these exercises take substantially longer than those with a fixed scenario, they better reflect reality and illuminate many challenges. For example, in addition to facility-specific challenges, numerous deficiencies were identified in the city's systems, including those relating to receiving patients, communications, and IPC measures.

The review meetings that followed these exercises became an opportunity to enhance face-to-face networking among relevant stakeholders. They addressed the identified gaps by clarifying their roles and responsibilities and discussing the need to prepare necessary human and other resources in advance of an emergency. Participants also agreed to further assess the problems identified in the exercise and to pursue collaborative solutions, in turn, improving knowledge and skills and raising awareness on crisis management among all relevant staff. Importantly, this collaboration is not only useful for addressing threats posed by emerging infectious diseases, but also for other events such as natural disasters and biological terrorism events.

Finally, these exercises have also prepared the hospitals to play a larger role in the broader local health system. For instance, while the three hospitals were not designated hospitals for highly infectious and virulent diseases, thanks to the exercises, these facilities all played a role in the local response to COVID-19 and accepted COVID-19 patients without significant trouble.

Conclusion

The risk of emerging infectious diseases in Japan and around the world is increasing due to urbanization and highly global travel networks. This chapter described recent efforts to reinforce public health preparedness and response capacity for emerging infectious diseases in Kawasaki City, a large city in Japan. Medical institutions are at the forefront of local infectious disease control and their cooperation with public health authorities is crucial for disease surveillance and the early detection of public health emergencies. Kawasaki City has implemented several actions to better prepare for pandemics, including developing the KIDSS, which has facilitated rapid information sharing between medical institutions and the health departments in the city; fostering trustworthy, multidisciplinary relationships through various forms of joint exercises that have been conducted over the last several years; and working to develop human resources for health — one of the most important response capacities — by establishing a development program through collaborations with national programs — a model that serves as a best practice for developing local human resources.

The COVID-19 pandemic has reminded us of the importance of preparing for public health emergencies before action is necessitated by reality. While Kawasaki City's previous efforts prepared it to respond and could serve as a model for other large cities, pandemic preparedness is a continuous cycle and the local-level preparedness and response efforts for the COVID-19 pandemic will be reviewed to determine what was effective and priorities for future work in preparing the city for infectious disease outbreaks.

References

[1] Government of Japan. Act on the prevention of infectious diseases and medical care for patients with infectious diseases (the Infectious Diseases Control Law). 1998. www.japaneselawtranslation.go.jp/law/detail/?vm=04&re=01&id=2830. [Accessed 14 June 2020].

[2] Government of Japan. Act on special measures for pandemic influenza and new infectious diseases preparedness and response. 2013. https://www.mhlw.go.jp/web/t_doc?dataId=78ab2871&dataType=0&pageNo=1. [Accessed 8 November 2020].

[3] Kawasaki City Infectious Disease Surveillance Center. Kawasaki City Infectious Disease Surveillance System (KIDSS). 2020. https://kidss.city.kawasaki.jp/en/. [Accessed 12 August 2020].

[4] Ministry of Health, Labour, and Welfare. The report of review meeting on the novel influenza (A/H1N1) countermeasures. 2010. www.mhlw.go.jp/bunya/kenkou/kekkaku-kansenshou04/dl/infu100610-00.pdf. [Accessed 14 June 2020].

Chapter 7

Making the case for biopreparedness in frontline hospitals: a Phoenix case study

Saskia Popescu

Schar School of Policy and Government - Biodefense, George Mason University, Arlington, VA, United States

Hospitals and healthcare facilities are the frontlines for identification, isolation, and defense against infectious disease outbreaks. Improper or substandard infection prevention and control (IPC) efforts directly facilitate the spread of disease within a community through delayed identification, isolation, and treatment of the patient, which prevents downstream public health efforts from responding to the case, not to mention the risk of transmission to staff. Infection control failures in hospitals were responsible for 31% of global Middle East Respiratory Syndrome Coronavirus (MERS-CoV) cases since December 2016 [1]. The two nosocomial (healthcare-associated) cases of Ebola virus disease (EVD) in nurses caring for a patient in Dallas, Texas, were the result of poor infection control practices and administrative support. When the first imported case of Ebola was identified within a hospital in the United States of America (U.S.), the spread of the disease to healthcare workers revealed substandard infection control practices that were common within the U.S. healthcare system. More recently, the novel coronavirus, SARS-CoV-2, which causes COVID-19, has shown a propensity for transmission in healthcare settings. Many states, like New York, California, and Washington have been severely impacted by the virus. One in particular, Arizona has been especially hard hit with COVID-19, proving to be one of the hotspots internationally with a 25% positive rate and a seriously strained healthcare system during a rapid surge in June and July 2020 [2]. Arizona's experience as a COVID-19 hotspot emphasizes the importance of this topic and sets the stage for this chapter's focus on a hospital system within Phoenix, AZ, and its efforts to prepare for biological events well before the pandemic. From over 90,000 healthcare worker infections worldwide, or the rampant transmission of the disease throughout long-term care facilities, COVID-19 is the latest example of how healthcare represents a unique vulnerability in the United States' biodefense [3].

Inoculating Cities. https://doi.org/10.1016/B978-0-12-820204-3.00007-3

115

Infection control failures also result in harmful economic outcomes, as was seen in the healthcare-associated cases of Severe Acute Respiratory Syndrome coronavirus (SARS-CoV) in Toronto, Ebola in Dallas, Texas, and MERS-CoV in Seoul, South Korea. The 2003 SARS-CoV outbreak spread throughout Toronto healthcare facilities and resulted in a $1 billion price-tag for quarantine efforts that were attempted as a control measure but ultimately cost in terms of tourism and domestic economy [4,5]. Treating an Ebola patients in the United States is estimated to have cost $1.16 million for two patients; and this figure does not account for nosocomial transmission and IPC failures that required additional contact-tracing and widespread environmental disinfection [6]. The inadequate IPC practices in Dallas, TX, resulted in two nurses becoming infected, a lawsuit, and severe financial loss for the hospital. In fact, the lawsuit states that the nurses at the Dallas hospital reported that the hospital administration failed to follow basic IPC principles, provide adequate training or personal protective equipment (PPE), and even resisted nursing demands for proper isolation of the patient [7]. Despite spending $60 billion on biodefense efforts since 2001 to protect against high-consequence diseases [8], a single Ebola patient resulted in significant infection prevention and control failures, economic damage to the hospital, lawsuits from staff, and fear within the United States [9]. The United States has the resources and capabilities to adequately defend against infectious disease threats and yet a single cluster of Ebola cases exposed a wholly vulnerable and deeply compromised level of healthcare biopreparedness.

The healthcare industry within the United States is unique, as it relies on primarily private entities. This is a unique approach to healthcare in industrialized countries, like the Bismarck model in Germany or the National Health Insurance model of Canada. Countries like the United Kingdom utilize the Beveridge model, which provides healthcare for all citizens and is financed by the government through taxes [10]. The United States is unique in that it spends a considerable amount on healthcare per person and has elements of several different models. In their 2020 Hospital Statistics, the American Hospital Association reported that there are 6146 hospitals in the United States, of which 5198 are community hospitals and 209 federal hospitals [11]. Moreover, there are 924,107 staffed beds across all US hospitals and over 36 million admissions with a total expense for all US hospitals exceeding $1.1 trillion. As the majority of US hospitals are private entities and ultimately determine their own level of biopreparedness, this creates a fractured approach to healthcare readiness for biological threats. A 2018 report by the Office Inspector General (OIG) of the U.S. Department of Health and Human Services assessed hospital readiness to emerging infectious disease threats following the 2013−16 Ebola outbreak [12]. The surveyed hospital administrators reported a higher level of preparedness but noted that emergency preparedness personnel often lacked the specialized knowledge required for infectious disease threats. Moreover, administrators noted that it was difficult to integrate procedures specific to emerging infectious diseases, specifically IPC, into emergency plans and that responding to such events was out of the comfort zone for emergency preparedness coordinators. As a result of these vulnerabilities and to better ensure a prepared healthcare system, the US sought to build a tiered approach at a national level.

The current tiered system in response to special pathogens

The eleven patients treated for Ebola in the United States revealed the challenges and risks associated with providing medical care to patients infected with special pathogens. Advanced medical settings translate to more invasive medical practices, which inherently puts healthcare workers at risk. In late September 2014, a traveler from Liberia would change the face of American healthcare preparedness. On September 20, 2014, Thomas Duncan traveled from Liberia where he was exposed to Ebola virus disease, to Dallas, Texas and would later seek treatment on September 25, 2020 [13]. As a result of treating Duncan, US hospitals went into overdrive to meet the needs and expectations of identifying and caring for a potential Ebola patient. These efforts were exhausting in terms of staff and resources, and fundamentally the level of preparedness was considered unsustainable in the long-term. Between the costs of waste management, extremely high PPE needs, or staff-to-patient ratio necessary, it quickly became apparent that the existing processes were not sustainable. In response, the Department of Health and Human Services was provided funds to develop a regional approach to hospital management of patients with Ebola virus disease. In collaboration with the United States Centers for Disease Control and Prevention (CDC), they developed a state and jurisdictional-based tiered hospital approach. Hospitals in the higher tiers receive funding through the Hospital Preparedness Program. These included frontline facilities, Ebola assessment hospitals, and Ebola treatment centers [14]. Frontline facilities are acute care hospitals — including emergency departments and urgent care clinics — and their guidance is to be able to rapidly identify and triage a patient with relevant exposure and signs and symptoms. The name was quickly expanded to refer to Ebola and special pathogens treatment centers. Frontline facilities are expected to be able to properly isolate suspected patients and patients under investigation (PUI), potentially coordinate with state and local public health authorities for testing, and manage the patients for roughly 12—24 h until they could be transferred to an Ebola assessment or treatment facility.

The next tier, Ebola assessment hospitals, are acute care hospitals that are prepared to receive and isolate a patient under investigation and care for them until an Ebola diagnosis can be confirmed or ruled out. These hospitals are expected to have the diagnostic capabilities to transport specimens, staffing for the patient, proper PPE, waste management, and infection control training.

Ebola assessment hospitals are expected to provide up to 96 h of evaluation and care for patients under investigation. If a patient is found to have Ebola, the assessment hospital is expected to prepare transport for the patient to an Ebola treatment center. Fifty-five hospitals were initially designated as Ebola treatment centers, meaning that they had the capabilities to treat patients confirmed to be infected with Ebola. Ebola treatment centers are expected to have a multidisciplinary team of clinical and non-clinical members ready to respond, ongoing training, laboratory protocols, and a designated space for possible point-of-care testing.

In 2015, the Office of the Assistant Secretary for Preparedness and Response, expanded the special pathogen hospital tiers to also include a regional network tier. This revision designated 9 of the 64 Ebola Treatment Centers — those with the greatest capacity — as Ebola and Other Special Pathogens Centers [15]. In 2017, an additional hospital was elevated to this tier, leaving 63 hospitals as Ebola Treatment Centers. The Special Pathogens Centers are expected to maintain competencies and capabilities to accept a suspected or diagnosed patient within 8 h of notification, as well as conducting drills quarterly. For hospitals, the voluntary designation as an Ebola treatment center is prestigious, but also comes with substantial cost and ultimately, public association with highly infectious diseases.

The goal of the tiered system approach was to alleviate the burden of high-consequence pathogen readiness for US hospitals and to create a specialized framework to ensure that all hospitals had a basic level of preparedness, while a select few were specialized and could fully manage infected patients. The considerable risk for transmission during hospitalization and medical care, especially in more advanced healthcare settings, was a serious gap identified in the 2013—16 Ebola outbreak [15]. Much of the literature does note that these failures would likely happen again, as preparedness for high-consequence pathogens is not a hospital norm [12,16].

However, there are inherent gaps in this approach. First, frontline hospitals, which represent the majority of hospitals in the United States, are widely neglected and knowledge is centralized in those higher tiers. Funding prioritizes state and regional treatment centers and assessment hospitals, with the expectation being that frontline hospitals manage their own efforts. As the aforementioned OIG report reveals, special pathogen preparedness does not represent a key initiative for many frontline healthcare facilities. The second issue relates to the volume of hospital beds within the biocontainment units. Most facilities report 2—4 beds, which equates to roughly 300 beds across the United States. Should there be a large-scale event that requires more hospital capacity for patients with special pathogens, this will become a significant roadblock to care. Lastly, there is a considerable challenge in financing these hospitals. Frontline hospitals must provide their open support for special pathogens efforts. From education to training, this falls upon them to prioritize and make a financial investment in high-consequence disease preparedness.

These efforts are not insignificant financial investments and the work around the 2013−16 Ebola outbreak sheds light on such monetary hurdles. While there is little information on the actual costs related to US hospital response to Ebola, there is even less for the hospitals that were considered frontline. Some estimates suggest the average amount spent by hospitals on combined supply and overtime labor costs was $80,461 (n = 133; 95% confidence interval, $56,502 − $104,419) and that small hospitals spent larger amounts on staff overtime costs per 100 beds than large hospitals [17]. Building a biocontainment unit is quite expensive as well. The mean cost to build such a unit is roughly $1.2 million, which likely means many hospitals are dependent upon federal funding for developing these additional measures [14]. Unfortunately, the funding for assessment hospitals and treatment centers expired in May of 2020, with the future of federal funding for every tiered hospital outside of the ten regional facilities, in limbo [18].

While the tiered hospital approach to special pathogens is a step in the right direction, there are considerable gaps in US hospital biopreparedness that ultimately rely on hospital administrators to prioritize readiness. Maintaining biopreparedness across US hospitals is complex as there are few regulatory requirements that require hospitals to invest in emergency preparedness and no requirements related to infectious diseases. The resulting situation is one in which hospitals can meet regulatory guidelines without investing a single dollar in infectious disease preparedness. While hospitals will prepare for influenza season by purchasing additional personal protective equipment and vaccines, unless hospital administrators view biopreparedness as a worthwhile investment, such endeavors for other pathogens are unlikely to be pursued. Maintaining preparedness for low probability but high consequence diseases, like Ebola or SARS-CoV, can be costly and with many competing priorities, it is not surprising that many facilities report dwindling attention to such biological threats.

Maintaining preparedness − a vulnerability

The 2018 OIG report also points to the challenges of maintaining readiness as competing priorities do little to encourage healthcare administrators to continually invest in costly preparedness resources. Even hospitals that opt to invest in biopreparedness report continued opportunities for improvement. A recent assessment of the New York City hospital emergency department's response to patients potentially infected with highly contagious or special pathogens revealed considerable gaps, even among hospitals classified as Ebola treatment centers. From December 2015 through May 2016, 49 New York City hospitals participated in 95 drills regarding measles and MERS-CoV scenarios. 39% of hospitals failed at least one drill and there were considerable infection prevention failures. Patients were appropriately masked and isolated in 78% of drills and infection control measures occurred more frequently when travel history was obtained (88% vs. 21%) [19]. The focus on travel history as a precursor for initiating masking and isolation reveals a major weakness: a dependence on international travel to consider advanced isolation for patients, which requires the healthcare worker to ask and the patient to be forthcoming. Ultimately, they found that infection prevention practices were poor, with correct hand hygiene performed 36% of the time, and only 16% of staff showing patients how to perform hand hygiene. Moreover, in 80% of the drills that required patient isolation, the proper isolation signage was only posted on the patient's room 70% of the time and staff wore the correct PPE 74% of the time [19]. These failures indicate not only suboptimal infection control compliance, but point to a continued lack of preparedness for infectious disease threats, whether they be rare or more common infections. The internal priorities of each hospital administration likely vary, but researchers have found that a majority of IPC programs reported a lack of administrative commitment to additional resources once an outbreak has ended [16]. While there are now designated hospitals to treat patients with special pathogens and many infection preventionists report that their facilities are more prepared, this survey was completed in 2015 and it is questionable if hospitals are still maintaining such efforts.

Addressing this vulnerability through a six-hospital system in Phoenix, AZ

Despite this stark reality, in 2018, a Phoenix-based healthcare system opted to invest in high-consequence disease preparedness. From performing a gap analysis to the creation of a subcommittee with key stakeholders, this was an endeavor that required considerable administrative support and financial investment.

Nestled across the metropolitan area of Phoenix, Arizona, exists a five-, soon to be six-hospital system with over 100 outpatient clinics and offices. The second largest hospital system in the state and the city of Phoenix, it includes three level-one trauma centers, representing a significant portion of the healthcare resources in Phoenix. In spite of the considerable investment during the Ebola response from 2014 to 2016, general biopreparedness had grown stagnant and there were no exercises planned for the immediate future.

In late 2018 though, following a shift in leadership and staffing within the IPC program, there was a renewed interest in assessing and improving overall infectious disease preparedness within the hospital system. All of the hospitals within the system fell in the category of frontline hospitals within the tiered special pathogens hospital program, meaning that the expectation was that they should be able to identify, isolate, and inform of any suspected patients with a special pathogen. Moreover, the expectation was these hospitals would have the capability to care for such patients for 24 h. The hospital system is also categorized as a non-profit, private health system and is not affiliated with state or federal hospitals. There are roughly 1340 in-patient beds across the system, including multiple intensive care units, a neonatal intensive care unit, several oncology units, and a military partnership for emergency preparedness. Early on, it was decided that investing in biopreparedness also meant investing in infection prevention efforts, as enhanced measures build upon foundational practices.

Gap analysis

The first step in addressing vulnerabilities was to establish a gap analysis to understand the existing state of readiness within the hospital system. Staff had not received training for high-consequence disease response since 2014 and that training was specific to Ebola. Since this was the most recent training and education provided, it was decided that Ebola would be used as the test-scenario in the gap analysis. The analysis included questions based on guidance from the CDC for the management of PUI as well as algorithms created by the Arizona Department of Health Services in 2014.

While relatively broad, the gap analysis aimed to assess processes and knowledge starting in the Emergency Department admission and triage process. The survey of staff mimicked what the admission of a PUI and the role of various departments and stakeholders. Staff within the emergency department were asked a range of questions that tested their knowledge of response algorithms, communication, which patient rooms were to be used, PPE, and waste management, among other relevant processes. Those surveyed included emergency department staff, nurses, and physicians, laboratory personnel, environmental services, materials management, security, facilities, IPC specialists, and also included a review of the electronic medical records system.

The gap analysis involved 44 components and ultimately identified several considerable gaps within the system [20]. The emergency department staff represented the biggest vulnerability. Staff were questioned on how they would respond to a febrile patient with a travel history to an Ebola-affected region. While there was a travel screening question asked in triage, staff did not commonly ask about recent international travel if the patient presented with symptoms that were not consistent with an infectious etiology. Intake staff within the emergency departments were aware of the importance of documenting travel history but no staff at any of the facilities could speak to how they would respond if a patient had a relevant travel history and symptoms that could indicate EVD or another high-consequence pathogen. Of those surveyed staff, 60% were able to vocalize a proper communication strategy but only 20% noted that the process involved contacting an infection preventionist. Staff were aware of an algorithm for response but could not speak to its location. Across all staff, there lacked an ability or comfort in skills to don and doff the appropriate PPE and all requested additional training. A majority of staff were able to indicate the designated rooms for patients under investigation, but only 80% were airborne infection isolate room (AIIR) or negative pressure, and all spaces required the construction of an ante-room or containment zone to block off the area and allow for enough space for PPE donning, doffing, and patient care within a designated, sectioned area. Only one hospital had a non-negative pressure room designated for a patient under investigation and staff vocalized that this was chosen due to the direct route from triage to the room, thus minimizing exposure during transport.

In terms of personal protective equipment, each hospital had been provided with enhanced equipment for the Ebola response in 2014, in the form of a cart that was a repurposed code cart. This cart included the necessary PPE, directions for donning/doffing, and overall guidance for patient assessment. When evaluated, none of the hospital staff could vocalize the location of the PPE, with 80% of the carts being held within a locked room in the basement location of materials management. There was no clear ownership of the cart and responsibility for ensuring it was stocked with non-expired goods represented a consistent administrative failure. Each party thought it was managed by another department, which rendered the carts inaccessible, out of date, and often with limited supplies.

Beyond the emergency departments, medical management for PUIs represented a significant hurdle as there was no specific plan for the care of patients and treatment was dependent on healthcare workers volunteering. Moreover, there were no plans for the transfer of a PUI from the hospital to another should they require a higher level of care. All guidance related to Ebola response outside of the designated carts was in binders within the IPC department and had not been reviewed since 2014. Fortunately, the systems for acquiring additional PPE within 24 h were in place. Interestingly, the existing waste management containers and contracts for Ebola-associated waste were considered to still viable. However, upon further review, the contracts were no longer valid. Laboratory staff were able to communicate the process and materials necessary for shipping samples to the CDC or state laboratory, as well as point-of-care testing equipment for use within the patient's room. There were no specific plans to incorporate security services into the process to ensure limited access to PUIs and the care area. Lastly, a survey of infection preventionists revealed that a majority felt they could mostly manage PUIs but required a refresher of the materials, as they had not reviewed processes for several years. Moreover, the infection preventionists all asked for additional training to ensure they were effective as subject matter experts during the care of PUIs.

Overall, the gap analysis found considerable vulnerabilities but none that were particularly surprising given the time that had passed since 2014 efforts. The next phase though would prove to be the most challenging — establishing a network-wide subcommittee and attacking the list of opportunities.

Development of a high-consequence disease subcommittee

Following the findings of the gap analysis, it was determined that to respond to each opportunity, a large, multi-disciplinary high-consequence disease (HCD) subcommittee would be established across the healthcare system. Ultimately, without the support of hospital leadership, this would not have been possible. Operationalized and managed by the IPC program, members/key stakeholders included infection prevention, laboratory, medicine, nursing, human resources, information technology (IT), environmental services, facilities and environmental controls, education, emergency department and critical care, supply chain, and emergency preparedness.

The subcommittee began meetings on a monthly basis and presented both the gap analysis but also the objectives and goals — to enhance preparedness for high-consequence diseases across the hospital system. Based on quarterly goals, these included addressing vulnerabilities identified within the gap analysis, creating educational and training sessions for frontline staff, establishing a drill timeline, and building PPE kits for each healthcare campus. It was determined early on, that while there was a critical and timely need to educate frontline staff, the subcommittee's efforts would focus on sustainable preparedness as a key goal.

The first priority within the HCD subcommittee was to communicate to key stakeholders what high-consequence diseases were and why they posed a risk to the continuity of care within the healthcare system. Providing the findings of the gap analysis allowed stakeholders to have a role and responsibility within the subcommittee and ultimately, deliverable action items. The HCD committee established goals and monthly meetings that focused on specific topics within the readiness of the healthcare system and having stakeholders report on their efforts (e.g., facilities and the testing of barriers to create anteroom-containments outside of designated HCD patient rooms for the doffing of PPE). Each meeting focused on providing updates regarding work completed and highlighted on a specific stakeholder's work to address the vulnerabilities.

In conjunction with this work, one of the infection preventionists was identified as the point person for supporting such efforts based upon their background in biopreparedness. From this, a HCD plan was developed as a point of reference to incorporate guidance, resources, and processes should a patient with a high-consequence disease be treated. The plan focused on two key approaches to HCD readiness: the i3 strategy (identify, isolate, and inform) and presenting three potential forms of high-consequence pathogens (i.e., viral hemorrhagic fevers, airborne organisms, and Disease X). The presentation of the three high-consequence disease categories was intended to help guide response efforts for staff based on disease transmission mechanisms and the respective response efforts required. Moreover, the inclusion of Disease X was an effort to acknowledge that there will likely be an emerging infectious disease in the future with a response that will be characterized by evolving guidance and response measures. The inclusion of Disease X in this strategy proved to be quite helpful during the early weeks of the COVID-19 outbreak.

The HCD plan was created to document the process changes identified within the gap analysis, but also to act as a singular reference for response efforts. The plan was made available online within an internal webpage and included a range of information, from CDC guidance on waste removal and disposal, to the steps required for donning and doffing PPE, lists of individuals to include in notifications, a reference guide, and lists of the designated HCD rooms within each emergency department and what the anteroom-containment barriers looked like. The goal of the plan was to ensure all the CDC-supported guidance was available in a single document to avoid confusion, delay, and missteps. Guidance documents from the Occupational Safety and Health Administration and the National Ebola Training and Education Center were also included and enhanced the HCD plan by ensuring its utility and applicability to the health systems. Moreover, pieces from each key stakeholder (e.g., laboratories) were updated to match the internal processes and provide detailed guidance that was relevant to the hospital system, compared to overarching, non-specific guidance from CDC. Lastly, the plan also established a goal of creating a Provider Response Team, which would be a voluntary team of nurses, physicians, and respiratory therapists who would undertake additional training and provide the majority of care for PUI to relieve frontline staff. One of the most important goals within the plan and the subcommittee was to provide HCD training (i.e., an overview of HCDs, internal processes, enhanced PPE training) for over 90% of the emergency department staff and urgent care providers (i.e., nurses, physicians, and respiratory therapists). The decision was made to focus the PPE section of the HCD training on enhanced PPE for Ebola and other viral hemorrhagic fevers, as it was the most complicated and utilized the least by healthcare workers. Isolation precautions for airborne diseases, like MERS-CoV, rely on PPE and isolation precautions that staff are more comfortable with, such as N95 masks and airborne isolate rooms. Prior to initiating this training though, it was critical to establish up-to-date PPE kits and ensure there was a continued supply chain.

Enhanced personal protective equipment kits

One of the first tasks of the HCD subcommittee was to ensure the supply of enhanced PPE as existing stocks were incomplete, inaccessible, or outdated. The infection preventionist HCD coordinator worked with the system-level supply chain to establish a list of necessary PPE and the volume that should be made accessible to each hospital campus for both training and actual patient care.

Surprisingly, this task proved to be challenging as much of the attention to Ebola enhanced PPE had diminished over the years and there was no longer such a keen interest from manufacturers. There were considerable challenges associated with finding appropriate surgical hoods, testing full-body suits versus gowns, and identifying the most universal-fitting, fluid impermeable boot covers. These challenges proved to be time-consuming and resulted in delays. Based on financial stewardship and training capacity for staff, it was decided to use N95 masks instead of powered air-purifying respirators. The identification, testing, and acquisition of equipment for enhanced PPE kits proved one of the more challenging endeavors despite considerable support from the hospital's supply chain director. In the end, it took over six months to identify the necessary PPE materials so that they could be built into kits for each campus. It was determined that each hospital would receive 60 kits — with 5 readily kits available in both the emergency and urgent care departments.

Harnessing the power of the electronic medical records

During this time, efforts to utilize electronic medical records (EMR) were made a priority. A sub-group of the HCD subcommittee, including members from IPC and IT, worked to refine an existing tool within the EMR that inquired about international travel upon admission. The work of this group allowed for modifications based on ongoing outbreaks and a better, more real-time alert for medical providers. Alerts were created in coordination with the EMR team and IPC, and were alerted providers of relevant outbreaks going on around the world. These alerts, known as a Best Practice Advisory (BPA) drew from CDC travel alerts determined to be Level 2 or higher. Upon answering a question regarding the history of international travel within 28 days, admitting staff would be prompted to indicate where the patient had traveled. If they noted that the patient traveled to an area with an existing alert (e.g., Wuhan City — COVID-19), a BPA would appear and alert the medical provider. Staff could then choose to order isolation precautions or acknowledge the BPA, which would then mute it. If the BPA was acknowledged, it would no longer alert that provider but would continue to alert others. Further, should the patient become febrile during their admission, it would re-prompt the BPA for medical providers. Within the BPA, it was noted to communicate with IPC and provided their contact information. The BPA alerts though were dependent upon the admitting provider indicating which country the patient had traveled to, which required additional education and communication efforts to ensure this step was not missed. The goal of the alert was to notify staff of existing outbreaks that they might consider both infection prevention efforts, but also diagnostics. Should a patient have a travel history to an area of an outbreak, the hope was that the BPA would encourage healthcare workers to isolate the patient to determine symptoms and risk for infection.

Environmental controls

A core component to preparedness and response for infectious disease events in a healthcare setting is that of engineering controls. The ability to build the anteroom-containments outside of designated HCD rooms and ensuring negative pressure is an easily forgotten task. One of the first tasks within the subcommittee was to ensure that all AIIR were functioning and maintaining true negative pressure, which required evaluation by a third-party vendor. Such efforts across multiple hospitals are neither easy nor inexpensive. Fortunately, the utility of AIIR extends beyond HCD efforts, and making the case for their evaluation was not difficult. Each hospital had at least one AIIR within the emergency department that would serve as the location for the care of a HCD patient with an airborne infection. However, if a patient was under investigation for a viral hemorrhagic fever, the room might change based upon the need for an anteroom-containment and other unique patient care components like a private bathroom.

A second component relating to environmental controls was the ability to rapidly build anteroom-containments outside of the designated HCD rooms. Those specific rooms were chosen in conjunction with emergency department directors, IPC specialists, and facilities, to ensure the best space was selected based on negative pressure, private bathrooms, continuity of care within the emergency department, and ease of patient transfer. As not all AIIR were designed to have private bathrooms, it was decided that should a patient with a viral hemorrhagic fever present to the hospital, it was ideal to have a private bathroom versus negative pressure, as the risk for healthcare workers was higher in removing human waste from a bedside commode than of aerosol-generating procedures. Such decisions required the consideration of varying risk categories and were not made lightly but represent the limitations that exist in many frontline hospitals. Ultimately, it was decided to work toward converting these designated rooms to negative pressure within the next year.

Once the designated room was chosen for a potential viral hemorrhagic fever patient, the infection preventionists at each hospital worked with the emergency department director and facilities to establish drills to build an anteroom-containment outside the space. These tests were conducted to ensure that materials were available, determine the time required for construction, and take photographs for inclusion into the plans as a source of reference. Coined "the barrier build," these drills acted as valuable teaching moments for not only the staff in the emergency department but also those facilities engineers helping to build the anteroom-containment. As each hospital and emergency department is unique, it was important to account for the variances that might impact patient care and staff safety.

Network-wide communication

During the establishment of the HCD subcommittee, it was determined that network-wide communication of their efforts should be provided to all facilities in the system. To this end, a logo was developed by the communications and marketing departments to help staff identify efforts that were associated with the HCD group. Further, on several occasions, the infection preventionists at each hospital campus would present at campus-wide meetings on the subcommittee, its efforts, and upcoming training opportunities.

Network-wide awareness of the HCD efforts was an important factor in not only ensuring support, but also for ensuring awareness among thousands of employees. Moreover, this provided an opportunity to discuss the PPE that was selected and ordered, and the approach of having a small "grab and go" supply in the emergency department with additional kits available elsewhere. Additional communication was provided by the IPC network director to hospital leadership as a means to fostering support for staffing training efforts. Updates on the HCD subcommittee were also provided at the bi-monthly meetings of the IPC committee.

Training — where to begin?

At a network level there was a desperate need to address HCD education and training but ambiguity about where to begin these trainings. The goal was to ensure all frontline staff had experience with enhanced Ebola PPE, could speak to relevant processes (e.g., i3), and locate additional materials. Thus, it was decided that instructional lectures would not only discuss what HCDs were but also the differences in response measures and isolation precautions. It was also decided that trained infection preventionists acting as the instructors would discuss the support system for those caring for patients and the plan to incorporate response teams for relieving frontline staff. This was the time to reinforce patient safety measures while ensuring staff felt supported by both processes and the HCD subcommittee efforts. Additionally, after instruction, staff would spend time learning to don and doff enhanced Ebola PPE and also perform routine tasks while wearing PPE to experience the limitations in dexterity and challenges of common medical movements.

The training was developed by IPC specialists and based upon lessons learned from 2014 efforts, experiences at the CDC, and resources made available via the National Ebola Training and Education Center. Feedback on the training was provided from nursing educators and frontline staff. Continuing education credits were also made available through the hospital's nursing education program. Furthermore, it was decided that the training would be provided at each hospital campus and urgent care clinic, limited to 90 min (i.e., a 20 min lecture, 60 min for PPE donning/doffing, and 10 min for questions), and provided in the morning and evenings to account for morning and night shifts and to ensure that attendance would limit disruptions to healthcare services. The expectation was that infection preventionists would provide the training, as would nursing educators who attended training at the Center for Domestic Preparedness and went through additional training with IPC.

Perhaps the most challenging aspect of education and training was ensuring staff attendance. As this curriculum was not finalized until the fall of 2019, ensuring frontline staff attendance during a severe respiratory virus season was not ideal. There were challenges in ensuring staff attendance as it was not mandated from nursing leadership but rather encouraged. IPC had to work closely with nursing and medical leadership at each hospital to ensure support and administrative push to ensure 90% of frontline staff received training. Competing priorities and a heavy educational burden on nursing staff meant that HCD training was not a major priority for many. Moreover, since this training was not a regulatory requirement, it was often left to the discretion of the emergency department director to promote and enforce attendance.

An unanticipated real-world drill

In the middle of the HCD subcommittee efforts, a unique situation occurred — a patient presented with gastroenteritis and travel history to the Democratic Republic of the Congo, which at the time was experiencing an outbreak of Ebola. In the fall of 2019, a patient came presented to the emergency department reporting recent travel history and signs and symptoms consistent with the screening criteria recommended by the CDC. At this point on the high-consequence disease preparedness efforts, the subcommittee had been established, met a handful of times, and PPE kits were being ordered. The only operational aspect though was the BPA alert within the electronic medical record system.

Still, the patient was rapidly isolated and the IPC team was notified. While most of the PPE for Ebola was outdated or inaccessible, the ongoing work of the infection preventionists at that hospital campus meant that PPE kit samples provided to the hospital prior to purchasing were available for use. By luck, the two infection preventionists at the campus had recently been training on Ebola PPE donning and doffing steps because of attendance at a training at the CDC. The infection preventionists guided the facilities team to build an anteroom-containment around an AIIR within the emergency department, allowing staff to doff PPE inside the anteroom-containment according to CDC guide-lines. Over the course of the next 24 h, the patient was evaluated and cared for by a single nurse on 12-h shifts. The IPC team stayed with the medical pro-viders to ensure they had a PPE assistant to guide donning and doffing efforts. There were delays in establishing communication with the local and state public health entities and challenges in determining what laboratory testing would be provided both internally and externally to determine the clinical state of the patient. Overall, the challenges in identifying and isolating the patient ultimately did not prove to be the weak points in response, but rather the coordination with local public health authorities and determining what tests laboratory staff felt comfortable performing despite public health officials feeling the patient was of low-risk for Ebola. Ultimately, the patient was found to have a common diarrheal illness and public health authorities felt he was of low-risk due to the geographical area of his travel abroad and the work he was doing. Roughly 30 h after his admission, he was discharged from the hospital.

An after-action review of the event proved to be helpful in terms of opportunities for improvement. Several key lessons were learned, mostly with regard to laboratory and waste management considerations. The determination of what laboratory tests could be safely completed in the hospital versus what the health department requested proved disorienting. Moreover, there was generalized confusion across all sectors regarding what labs needed to be drawn for Ebola testing and the transportation process. As a result of these issues, extensive work was done with the hospital systems' laboratories to determine and establish protocols for what could be completed if the patient was of a greater risk category, but also establish point of care testing protocols. In terms of waste management, the previous contracts from 2014 were no longer viable and while the patient was ultimately determined not to be of risk for Ebola, had the waste required true decontamination efforts consistent Category A waste, it would have required an entirely new contract and thus a delay in waste remediation. Other smaller stresses were apparent during the event, such as information management and security around the area as paramedics and non-hospital employees in the emergency department sought to obtain information. While these were stressful lessons learned at the time, they proved extremely helpful for building better response and preparedness efforts.

COVID-19 — a test of existing efforts

In early 2020, news of a novel acute respiratory condition in Wuhan, China, began to prompt concern internationally. Within a matter of weeks, the first case of the disease now called COVID-19, caused by the novel coronavirus SARS-CoV-2, was identified in the United States. Hospitals across the country went on alert and worked to prepare for potential cases as tens of thousands were identified in the Chinese epicenter of Wuhan. Fortunately, the existing HCD infrastructure within the healthcare system had laid the groundwork for response efforts within the hospital system. Early on, there was network-wide communication regarding the outbreak, symptoms, i3 efforts, guidance for testing based on CDC criteria for PUI, and communication resources.

Within 2 days of the first case in the United States, the aforementioned communication was sent to staff and leadership. Infection preventionists at each campus were rounding in the emergency departments to ensure they were informed and prepared to identify and isolate potential patients. Daily nursing huddles ensured IPC efforts could provide instant updates and answer questions promptly. During this time there was concern regarding international PPE supplies, so the HCD subcommittee worked with supply chain stakeholders to acquire more masks and conduct daily assessments of supply levels. A new and enhanced BPA was created via the electronic medical record team to screen for not only travel to the relevant areas, but also send alerts that briefly explained the outbreak, the need for isolation precautions, symptoms, CDC criteria for testing, and instruction to call IPC immediately. This BPA was available for both inpatient and outpatient admitting documentation. If triggered, the BPA would allow providers three options — order isolation precautions, determine the patient did not meet criteria (i.e., no symptoms, etc.), or defer for chart review but would continue to alert until cleared. The EMR team was also able to tie the BPA alert to a text message and email that would be sent to the entire IPC team. This rapid alert ensured all cases were reviewed by an infection preventionist, but also that a more proactive response could be established, which helped frontline staff to feel more supported.

The following week and with the first COVID-19 case identified in Arizona, weekly meetings were established with an expanded HCD subcommittee attendee list. Meetings initially lasted 2 h to provide an overview of the situation and updates from each stakeholder (e.g., supply chain, updates on AIIR functionality, etc.), but these were reduced to 45 min to address weekly progress and needed resources. Shortly after it was announced that there was a case identified in Phoenix, AZ, several hospitals became inundated with not only close contacts of the case, but also worried individuals as well. This proved to be challenging for both frontline staff and IPC but was beneficial as it provided educational opportunities and opportunities to establish a better response. Several meetings were scheduled daily to establish working groups to address everything from communication to testing and surge capacity efforts. As community transmission became apparent in early March 2020, it was decided that more IPC rounding and education were required. Managing PPE supply strains also proved to be difficult and communication regarding the re-use and extended use of N95 masks was made a priority. Ultimately, it was found that the existing HCD subcommittee efforts and infrastructure helped lay the groundwork for COVID-19 response and as a result, attendance for the HCD training skyrocketed during this time.

The efforts of the HCD subcommittee would not be possible without the dedicated members, but also the support of the IPC director and those dedicated infection preventionists. While such efforts should not be dependent upon the passion and dedication of a select few, without their steadfast desire to enhance preparedness and response, these efforts would not be possible. The decision to focus resources and efforts on biopreparedness helped to ensure that a rapid system-wide response to COVID-19 was possible. The investment in hospital preparedness for biological threats should be seen as an investment in not only infection prevention and patient safety, but a reduction in vulnerabilities to unanticipated infectious disease events. As public health and healthcare are uniquely intertwined and intrinsically reliant on each other, these efforts helped not only strengthen this relationship, but reduce the burden when COVID-19 cases surged in the United States in June and July of 2020.

In truth, no hospital system is fully prepared for a pandemic and the nuances that come with it. COVID-19 was no different, but efforts to enhance readiness and reduce the strain in healthcare yielded benefits that trickled into public health. A hospital more prepared is one that can help reduce community cases by avoiding hospital-associated infections in patients and staff alike. Focus on enhancing laboratory capacity internally eased the burden of local labs and sped up patient notification, while focusing on PPE utilization and sustainable approaches helped keep healthcare workers safe and the hospital operational. Enhancing alerts and diagnostic capacity translated to faster reporting to local health departments, which helps speed up contact-tracing. All of these initiatives not only build internal readiness and response to biological events but also strengthen local public health and help reduce pressures on those critical resources.

References

[1] World Health Organization. WHO MERS-CoV global summary and assessment of risk. Geneva: WHO; 2017.

[2] Popescu S. Arizona reopened too fast. In: Epidemiologists knew it, but we couldn't stop it. The Washington Post; July 20, 2020.

[3] Mantovani C. Over 90,000 health workers infected with COVID-19 worldwide: nurses group. Reuters; May 06, 2020.

[4] Gupta AG, Moyer CA, Stern DT. The economic impact of quarantine: SARS in Toronto as a case study. J Infect 2005;50:386—93.

[5] Keogh-Brown MR, Smith RD. The economic impact of SARS: how does the reality match the predictions? Health Pol 2008;88:110—20.

[6] Sun LH. Cost to treat Ebola in the U.S.: $1.16 million for 2 patients. The Washington Post; November 18, 2014.

[7] Brown T. Dallas nurses say infection control ignored in Ebola care. Medscape; October 15, 2014.

[8] Hayden EC. Biodefence since 9/11: the price of protection. Nature 2011;477:150—2.

[9] Isidore C, Alesci C. Dallas hospital hit by Ebola losing patients and money. CNN Money; October 17, 2014.

[10] Wallace LS. A view of health care around the world. Ann Fam Med 2013;11:84.

[11] American Hospital Association. 2020 fast facts on U.S. Hospitals. Chicago: AHA; 2020.

[12] U.S. Department of Health and Human Services Office of Inspector General. Hospitals reported improved preparedness for emerging infectious diseases after the Ebola outbreak. U.S. Department of Health and Human Services; October 16, 2018.

[13] Berman M, Brown D. Thomas Duncan, the Texas Ebola patient, has died. The Washington Post; October 08, 2014.

[14] Herstein JJ, Biddinger PD, Kraft CS, Saiman L, Gibbs SG, Smith PW, et al. Initial costs of Ebola treatment centers in the United States. Emerg Infect Dis 2016;22:350−2.

[15] World Health Organization. Ebola health worker infections. 2020. www.who.int/features/ebola/health-care-worker/en. [Accessed 20 July 2020].

[16] Morgan DJ, Braun B, Milstone AM, Anderson D, Lautenbach E, Safdar N, et al. Lessons learned from hospital Ebola preparation. Infect Contr Hosp Epidemiol 2015;36:627−31.

[17] Smit MA, Rasinski KA, Braun BI, Kusek LL, Milstone AM, Morgan DJ, et al. Ebola preparedness resources for acute-care hospitals in the United States: a cross-sectional study of costs, benefits, and challenges. Infect Contr Hosp Epidemiol 2017;38:405−10.

[18] Popescu S. Outbreaks of lethal diseases like Ebola and the Wuhan coronavirus happen regularly. The US government just cut funding for the hospitals that deal with them. Bulletin of the Atomic Scientists; January 22, 2020.

[19] Foote MM, Styles TS, Quinn CL. Assessment of hospital emergency department response to potentially infectious diseases using unannounced mystery patient drills − New York City, 2016. MMWR 2017;66:945−9.

[20] Popescu S, Leach R. Identifying gaps in frontline healthcare facility high-consequence infectious disease preparedness. Health Secur 2019;17:117−23.

Chapter 8

Urban pandemic preparedness in Myanmar: leveraging vertical program capacities and the development of public health emergency operation centers

Elliot Brennan, Kyaw San Wai, Kyi Minn, Sandii Lwin
Myanmar Health & Development Consortium, Yangon, Myanmar

Myanmar, with a population of 54.75 million, sits between the world's two most populous countries, India and China, along with Bangladesh to its west, and Thailand and Laos on its eastern borders.[a] The country's central location among populous low- and middle-income countries (LMICs) representing 3.6 billion people or 47% of the world's population, as well as porous borders and contested areas, means its capacity to respond rapidly and effectively to health emergencies is both a regional and domestic imperative. This capacity has steadily improved over the past decade and more so under the civilian National League for Democracy government, which has emphasized health care and public health capacity. Yet, as several recent reports on pandemic preparedness note, Myanmar, like most countries, remains underprepared [1,2]. Moreover, Myanmar is at an early stage of urbanization compared to regional counterparts, with strong urban population growth expected in the coming years [3]. Such growth will significantly impact public health, particularly in urban environments. Prudent management and planning will be required, as will innovative solutions, to combat problems already experienced in other countries. Early indications suggest leveraging the strengths of various vertical programs and building on them to create public health emergency operations centers (PHEOCs) may serve as a viable model for cities in Myanmar and other low-resource settings.

---

a. The use of both Myanmar and Burma in this chapter aims to correctly use the name of the country at the referred time in order to better situate the reader and does not reflect any political statement.

Inoculating Cities. https://doi.org/10.1016/B978-0-12-820204-3.00008-5

Diseases have shaped the course of Myanmar's history. Several epidemics changed the nature of old Burmese kingdoms and helped shape the modern state [4]. For example, a 10th century cholera epidemic pushed people of Haripunjaya, encompassing modern-day northern Thailand and eastern Myanmar, to migrate into Lower Burma, changing the demographics and character of the Pagan kingdom [5]. Diseases also created difficulties for lowland coastal "Lower Burma" to project control over inland "Upper Burma" and highland border regions dominated by ethno-linguistic minorities. This was most pronounced during colonial rule, where the British centered in Rangoon (now Yangon) struggled to control Upper Burma and border regions amid public health challenges such as cholera, typhus, plague, and smallpox outbreaks [6].[b] Increased travel, seasonal labor movements, and permanent migration from rural locations to urban centers have increased the risk of community spread of communicable diseases and placed new stresses on urban health services. This contextual background and its modern-day impacts are important in helping to situate the discussion of public health emergencies and response in Myanmar.

The 2019 Global Health Security Index assessed Myanmar favorably for many of its global health security capabilities [1].[c] The index ranked countries based on six separate categories (prevention, detection, rapid response, health sector, compliance to norms, and risk environment). Myanmar scored favorably in detection and reporting, rapid response, and international norms compliance, but the report also noted a fragile health system, a high-risk profile relating to political and governance effectiveness, and low ability to prevent the emergence or release of pathogens. While not a member of the Global Health Security Agenda, Myanmar's compliance with international norms includes cross-border agreements with neighboring countries — such as a 2017 Health Cooperation Memorandum of Understanding with China under the Belt and Road Initiative, and its membership of the Mekong Basin Disease Surveillance Network and the Malaria Genomic Epidemiology Network. Moreover, Naypyitaw's 2017 completion of an International Health Regulation Joint External Evaluation (JEE) and a World Organisation for Animal Health's

---

b. An estimated 82.7% of British forces deployed in Burma during the First Anglo-Burmese War (1824–26) died from diseases, whereas only 4.2% were killed in action. For more, see: Robertson TC. *Political Incidents of the First Burmese War*. London: Richard Bentley; 1853.

c. COVID-19 exposed some of the challenges in assessing pandemic preparedness and the Global Health Security Index became fodder in the debate on what is effective preparedness. Indeed, many of the countries that performed strongly in this Index fared poorly during the first six months of the pandemic. While the ranking nature of the Index itself may be unhelpful, as noted by Razavi et al. [6a], the ambition and the information collected for the Index are commendable. While its findings need to be weighed with other evidence, as the authors themselves have argued, the conclusions on Myanmar presented here aim to illustrate one piece of the literature in health security. If nothing else, the Index demonstrates the difficulties of preparing for health emergencies — a country needs tremendous capacity and resources but also a good dash of luck.

Performance of Veterinary Services assessment demonstrated the country's commitment to improving emergency preparedness in line with international norms. Following these assessments, the Ministry of Health and Sports (MOHS) sought to build core capacities outlined in the International Health Regulations and developed a costed five-year, multi-sectoral National Action Plan for Health Security 2018−2022.

The 2003 SARS epidemic, along with the 2009 H1N1 influenza pandemic, and the 2019 coronavirus disease (COVID-19) pandemic have added an extra dimension to the ramifications of pandemics beyond public health − massive economic impacts that not only affect jobs and livelihoods, but also entire economies long after a pandemic subsides. As an emerging economy with large growth potential but with continual and systemic challenges, there is a strong economic argument for greater pandemic preparedness in Myanmar. This was reiterated by the health and economic impacts of the West African Ebola outbreak in 2015 that underscore the need to further strengthen health security capacity. The World Bank estimates that Guinea, Liberia, and Sierra Leone lost at least US$2.2 billion in forgone economic growth because of the epidemic [7]. A revised estimate places the economic toll of SARS around US$8.6 billion loss in Gross Domestic Product (GDP) for Hong Kong and Singapore and US$7.1 billion loss in foreign direct investment in China [8]. According to the Asian Development Bank, the global economic impact of COVID-19 could reach between US$5.8 trillion to US$8.8 trillion or between six and 10% of the global GDP [9]. Failure to prevent or mitigate a pandemic could trigger a series of cascades that can push segments of the population into poverty that will also lead to poorer health status and increased vulnerability to further shocks.

In the context of health security, an understanding of the country's particular environmental fragility is also important, affecting both urban and rural areas. According to the Global Climate Risk Index 2019, Myanmar ranked the third most climate change affected country in the world between 1998 and 2017, with Cyclone Nargis killing an estimated 140,000 people in May 2008 [10,11]. With the trend likely to continue in Myanmar, there has been an emphasis on emergency preparedness more generally. Growing international interest in One Health and impacts of environmental changes as a health determinant may support greater interaction between infrastructure designed for public health emergency response and those for natural disaster emergency response [12,13]. This can be seen in past and current planning documents. While geared largely toward natural disasters response, Myanmar's Interagency Emergency Response Plan supports pandemic preparedness through risk assessment, establishing minimum preparedness actions, standard operating procedures, and contingency plans [14]. This is supported by other documents such as the 2006 Pandemic Preparedness and Response Plan and the Early Warning, Alert and Response System that can support the deployment of Outbreak Control Teams in health emergencies

[15,16]. Moreover, and as discussed in this chapter, two emergency operation centers were established under two separate ministries.

The 2019 Monsoon season (i.e., June to October) witnessed a number of infectious disease outbreaks in Myanmar cities, including seasonal influenza, H1N1 and H3N1 influenza, along with vector-borne diseases such as Chikungunya and dengue fever — the last of which is currently undergoing an epidemiological shift and a major regional resurgence. These developments prompted the MOHS to fast-track the establishment of public health emergency operation centers in order to develop capacity to better respond to urban pandemics. More recently, the COVID-19 pandemic has highlighted the systemic weaknesses in Myanmar's healthcare system and reinforced the need to further develop and strengthen its public health emergency preparedness and response capacity. As the pandemic exacts its toll around the world, Myanmar (at the time of writing) had only reported a few hundred cases in the first half of 2020. Despite the health systems challenges, the MOHS stepped up its surveillance and information sharing efforts and activated its PHEOCs, guided by the World Health Organization's 2005 International Health Regulations. This chapter explores efforts at urban pandemic and public health emergencies preparedness in Myanmar, and discusses how broader urban pandemic preparedness can be strengthened by leveraging capacities that have been established through disease-specific programs, such as those for malaria and tuberculosis.

Urban pandemic preparedness context

Myanmar's approach to urban pandemic preparedness is built on three key factors: (i) its health status and low resource setting; (ii) Myanmar's urban environment; and (iii) the health security landscape in cities. The approach also draws heavily on experiences tackling emerging and re-emerging infectious diseases. These factors define the context through which the country has to assess public health and urban pandemic threats, and mobilize resources and expertize. While many of the pressing health security concerns are zoonotic diseases with origins in rural or remote areas, factors such as connectivity of such regions to their respective urban centers, weak surveillance and health systems, and widespread informal migration have become critical urban health challenges.

Health status and low resource setting

Myanmar and its urban centers experience challenges similar to those experienced in other low-resource settings. Despite significant progress in recent years, Myanmar has the lowest life expectancy at birth (64.7 years), second highest maternal mortality ratio (282 deaths per 100,000 live births), and high under-five and infant mortality rates (72 and 62 deaths per 1000 live births respectively) in the Association of Southeast Asian Nations (ASEAN). The under-five mortality rate is higher in rural areas, as are the rates of underweight and stunted children [17]. In early 2020, there were an estimated 600 critical care beds (including 330 intensive care unit beds) in the entire country, or approximately 1.1 beds per 100,000 population — one of the lowest distribution rates in the Asia-Pacific [18]. Recent work highlighted the differences between rural and urban populations in health-seeking behaviors for fever, which showed rural participants were less likely to seek or be treated by a trained provider than their urban counterparts [19].

Since social, economic, and political reforms began in 2011, the health sector has received significant investment, with government health spending climbing from MMK86 billion (US$98 million) in FY 2011–12 to MMK1.17 trillion (US$820 million, US$15 per person) for FY 2019–20, representing 3.32% of the government budget and 0.9% of the GDP [20,21]. Currently, around 74% of total health spending derives from out-of-pocket payments at the point of care, the highest in ASEAN [22]. Further, high out-of-pocket costs undermine health seeking behaviors that can allow diseases to take hold in communities, and result in patients only seeking treatment in advance stages. Financial barriers have also led to patients resorting to quacks, self-medication, traditional medicine or faith-healing that mask disease burdens and can unwittingly aid transmissions [23].

Recent budgetary increases have translated primarily into capital pro-curement — such as constructing primary healthcare facilities and diagnostic equipment [24] — as the health education system has seen delayed responses in expanding the quality and quantity of graduates. The National Health Plan (2017–21) reported 1.33 health workers (i.e., physicians, nurses, and other health staff) per 1,000 population in late 2016, with physician-to-patient ratios being far better in urban than rural areas [25]. Issues such as non-competitive wages and shortage of healthcare workers in remote rural areas have contributed to weak recruitment and retention of qualified human resources for health, leading to a systemic human resource shortage that affects the MOHS's capacity for overall planning, preparedness, and response to existing and potential public health challenges. Within the MOHS's health infrastructure expansion, urban specialist facilities have received far greater investment than primary care and rural facilities [26].

The urban environment in Myanmar

The 2014 census reported that 30% of Myanmar's population lived in urban environments, with the country having 14 towns and cities with populations greater than 100,000. Approximately 12.2% of the population resides in urban agglomerations larger than 1 million — namely Yangon (5.5 million), Mandalay (1.3 million), and Naypyitaw (1 million) [27,28]. This figure has steadily increased from 9.4% in 2000 and 10.9% in 2010. Many urban centers have witnessed significant growth in recent years as freedom of movement and economic prospects increased. And while poverty has been reduced significantly in recent years, 25% of the population remains below the official poverty line and a further 33% are highly vulnerable to falling into poverty [29]. Poverty was 6.7 times higher in rural areas than urban areas, with poor households having twice as many children on average [29].[d]

Urban population growth rates have been greatly impacted by the country's political and development situations. For example, Yangon's satellite towns were established by governments to address squatter problems and portray "national progress". North and South Okkalapa, and Thaketa were established by the military caretaker government in 1958, while over 300,000 people were relocated to peri-urban areas such as Shwepyithar, Hlaingtharyar, and Dagon Myothit by the State Law and Order Restoration Council junta in the late 1980s and early 1990s [30]. Cyclone Nargis' devastation of the Irrawaddy delta in 2008 later sent a surge of migrants to Yangon, with many settling in Hlaingtharyar. These areas continue to attract migrants from across the country — including an estimated 270,000 squatters and have also become synonymous with health challenges [31−33] — which can further strain urban health services.

The 2014 census reported a 30.4% annual population growth rate in Mandalay and a 14% annual growth rate in Yangon [34]. These exceptional rates were seen in the years immediately following Myanmar's political opening that also ushered new economic opportunities. This population explosion created significant housing problems, including short-term over-crowding and longer-term issues of rapid, poorly regulated construction that stretched the capacity of public services including those for health. While these high growth rates have now slowed to levels corresponding to those typical of a developing LMIC, the impacts of the "sugar rush" of the high-growth years are still being felt.

Increased urbanization has compounded the problems posed by under-investment in urban infrastructure, impacting social determinants of health. Parts of major cities in Myanmar have access to partially treated water supplies but smaller cities often rely on untreated supplies. Further, the piped sewage systems available in the three largest cities, draw from infrastructure from colonial times. In Yangon, for example, only one-third of the population has piped water while many rely on a sewerage system constructed in 1889 that was designed to serve a population 40,000 people [35]. Pollution and informal settlements are also on the rise and there are considerable infra-structure struggles to meet demand, leading to traffic congestion, flooding, and waste management challenges [27].

---

d. According to this report, the poverty line in 2017 was MMK1,590 per adult equivalent per day, people with consumption levels at or below MMK1,590 per day were considered poor.

Significant expenditure would be required to meet infrastructure demands, including basic requirements such as wastewater management, public transit, and livable neighborhoods alongside healthcare and social services. A 2013 McKinsey Global Institute Study found that Myanmar's major cities would require tens of billions of dollars to upgrade infrastructure to meet the needs of current and projected populations through 2030 [36].[e] Shortfalls in urban infrastructure can have an impact on high rates of urban diarrhea-related and vector-borne diseases, compounding the health challenges confronting vulnerable inhabitants.

Urban flush and water-sealed toilet usage have improved in recent years but remain insufficient in informal settlements in peri-urban areas. Solid waste management has also improved but capacity remains limited, leading to inadequate waste collection and the dumping or burning of solid waste, causing further health problems [36].

While it is easy to dwell on the many challenges confronting urban environments in Myanmar, it should be noted that considerable investments and improvements made in recent years by both government and civil society (e.g., a significant expansion of Yangon's garbage truck fleet and the Clean Yangon Campaign) have had positive effects on urban infrastructure that, in turn, will benefit urban and national health security. Furthermore, a well-being survey of residents in five cities — Yangon, Mandalay, Taunggyi, Mawlamyine, and Monywa — reported largely positive views about life in their cities, with the majority of respondents noting that they felt their city was heading in the right direction [37]. Urban respondents noted high social engagement, enjoying strong personal relationships, high levels of personal well-being and generally low levels of anxiety. These levels of general well-being among urban residents highlight the underlying strength of the social fabric and potential resilience to disaster or societal shocks, and a crucial component for collective response and ownership against pandemics. In general, social trust was higher in cities compared to other regions in the country.

Informal population hubs in Myanmar

While urban environments are often defined by both population size and more advanced infrastructure, there are many population hubs of well-over 100,000 population with little formal infrastructure. Many of these are due to crises or large economic migration. They too are a part of the rich tapestry of urban pandemic preparedness, some due to their mobility, others merely their sheer size, and most due to their vulnerability to disease outbreaks.

---

e. These estimates assume that Myanmar's large cities reach infrastructure levels in 2030 similar to what we see today in comparable cities in China, Indonesia, South Africa, and other developing countries.

In western Rakhine state, the parallel humanitarian crisis and internal conflict have resulted in large numbers of refugees now living in large over-crowded camps, along with internally displaced persons (IDPs) inside Myanmar residing mainly in tightly packed camps outside large cities and towns. Conflict in other states have also resulted in a string of IDP camps, though conditions are not as restrictive as in Rakhine. These camps require significant resources to support health and basic needs and provide a potential ground for the spread of infectious diseases, with many witnessing higher morbidity and mortality during epidemics.

This was on display during the early months of the COVID-19 crisis. In March 2020, tens of thousands of migrants from Myanmar rushed back across the main Thai-Myanmar border crossing at Myawaddy fleeing Thailand's lockdown and Myanmar border restrictions imposed in response to the pandemic. Others, still, returned through informal crossings. This over-whelmed border health checkpoints and significant concerns arose regarding the possibility of widespread community transmission as the migrants dispersed to cities, towns and villages across states and regions [38].[f] Such movements of IDPs and migrant workers may pose real risks for the spread of infectious disease across the country — especially as they head to densely populated urban environments.

Economic migrants intimately tie Myanmar's major cities and towns to the dense jungles where illegal resource extraction (e.g., jade and gold mining) is rife. Notably, the jade mines around the city of Hpakant in northern Myanmar have an estimated 300,000 migrants from all over Myanmar working in unsafe conditions overlapping with HIV, tuberculosis, malaria, and drug abuse hotspots [39]. Southeastern Myanmar meanwhile serves as the source for the largest number of Myanmar migrants in Thailand while concurrently its rubber and palm oil plantations, punctuated by dengue and malaria hotspots, attract domestic migrant workers mainly from central Myanmar [40].

Health security in key cities

Beyond its major cities and peri-urban areas, Myanmar's main border cities with China (Muse, Chinshwehaw and Mongla) and Thailand (Tachileik,

---

f. In Myanmar's administrative structure under the 2008 constitution, its 7 States and 7 Regions are parallel but equal entities with the ethnic majority Bamars (or Burmans) chiefly inhabiting the economically more developed regions and different ethnic minorities mainly living in the resource-rich States. The capital Naypyitaw is a Union Territory, while five self-administered zones and one self-administered division also exist. However, these divisions are not clear cut with many Bamar living in states (including both historical communities and recent economic migrants) and urban migration (both historical and recent) bringing many ethnic minorities to the major economic centers located in the regions, and in other states. The term "State/Region" is used to refer to institutions that exist in both State and Regions, of same rank and duty.

Myawaddy, and Kawthaung) and their surrounding economic catchment areas are arguably among the most important health security hotspots. These cities are the main arteries of both regulated and illegal trade between Myanmar and its two main trading partners that see significant documented and undocumented cross-border movement. Regional connectivity projects such as the East-West Economic Corridor and components of China's Belt and Road Initiative tie Myanmar to her neighbors through these border towns — that also serve as major economic hubs in their own right — attracting migrants from all parts of Myanmar. Complicating the health security landscape is the long-running internal conflict with ethnic armed groups (EAGs) that has placed many rural communities beyond the effective reach of the formal health sector. Although covered in varying capacities by EAG-administered ethnic health organizations (EHOs), these gaps have become major factors in the rise and spread of antimicrobial resistance across the border areas [41,42].

Myanmar's location in the Greater Mekong Sub-region (GMS) also places it in a zoonoses hotspot [13]. The country is considered a hub and transit point for the illegal wildlife trade in Southeast Asia and demand for wildlife from China making towns such as Mongla a major nexus in the GMS illegal wildlife trade [43]. This thriving trade has been complemented by live animal and poultry markets in urban centers across the country, with those in border areas also selling illegal wildlife or bush meats. Despite some discussion of relocating live animal markets from outside urban centers to their outskirts — likely driven by economic rather than health incentives — live animal and poultry markets remain in urban centers. For example, the "Chicken-Duck Market" remains within metropolitan Yangon's Mingalar Taung Nyunt township selling live poultry, while another township (i.e., Insein) has numerous pig slaughterhouses, most of which fail to follow hygiene regulations [44,45]. The 2017 announcement of relocating live-animal urban markets has not been fully enforced, nor have new sanitation standards been adopted in many wet markets. These factors lead to significant health security concerns in urban environments across Myanmar.

Experiences tackling emerging and re-emerging infectious diseases

The policy and technical expertise for Myanmar's pandemic preparedness draw heavily on the country's health status, low resource setting, and experiences tackling emerging and re-emerging infectious diseases. Furthermore, Myanmar is subject to a wide variety of natural disasters, particularly earthquakes, floods and cyclones, placing post-natural disaster health challenges such as water-borne diseases. While many of these natural disasters have their first impacts in rural or remote communities, their impacts are increasingly felt in urban environments through increases in food and commodity prices and affected populations migrating to cities.

Vertical programs and experiences in responding to HIV/AIDS, malaria, and tuberculosis

The country's pandemic preparedness capacity is supported in part from strong capacity in addressing communicable diseases. These efforts are sustained by significant funding and capacity building from international donors, centered around HIV/AIDS, tuberculosis, and malaria. While vertical, disease-specific funding has disadvantages — drawing staff from public service toward higher wages offered by non-governmental organizations, possibly hindering overall health systems strengthening, public sector human resources or altering aspects of quality of care — in Myanmar it has supported capacities relating to surveillance systems, laboratory systems, and the response to infectious disease outbreaks. Moreover, with a weak health system, particularly outside of major cities, this capacity cultivated by vertical or disease-specific programs is the foundation for pandemic preparedness. Examining the experience of such programs is therefore instructive.

In creating PHEOCs, the MOHS can leverage and learn from surveillance systems developed for vector-borne and infectious diseases to serve as the key resource for developing the PHEOCs for pandemic preparedness. The vector-borne disease control unit, especially the National Malaria Control Program (NMCP), is one of the stronger and more established surveillance systems in Myanmar. This is largely the result of concerns regarding the spread of multi-drug resistant malaria in the GMS — which has made malaria one of the most pressing regional health security challenges and resulted in the NMCP playing active and leading roles in health security, disease surveillance, and emergency preparedness. Neighboring Laos has used malaria as an entry point for developing PHEOCs for comprehensive pandemic preparedness and response [46]. Likewise, the National Tuberculosis Program (NTP) has accumulated surveillance and diagnosis expertize that is of crucial importance for other respiratory diseases.

Experiences responding to other public health emergencies of international concern in urban areas

Emergency management often fails due to a lack of leadership, coordination, integration, resources, planning, or proper communication in times of crisis. Few countries, even high-income ones, manage to perform consistently across all these areas. Several deficiencies were noted by Myanmar's 2017 JEE, exposed in previous crises and discussed in this chapter [47]. Lack of integration is evident in most countries with on-going conflict and a "decentralization by default" health system governance may be present [48,49]. From this, break-downs in leadership, coordination, planning, resources, and communication may follow. Recent health crises in Myanmar's urban centers have shown improvements in many of these areas with clearer communication to the public, more transparent and accountable leadership, and some improvements in resourcing. The National Health Plan (NHP) lays out clear pathways for improvements across the health system to increase these capacities.

In recent years, Myanmar's urban centers have responded to several Public Health Emergencies of International Concern (PHEICs) that are informing its response to the COVID-19 pandemic as well as future responses. By and large, these PHEICs have had a small imprint on Myanmar, at least to the extent they have been tracked and reported. The MOHS has instead focused on responding to numerous endemic communicable diseases, such as HIV/AIDS, tuberculosis, malaria, and other vector-borne diseases such as dengue and Chikungunya. The country has progressively improved its response to H1N1 since it was first declared a PHEIC in 2009. When alarm spread in Yangon of an H1N1 outbreak in mid-2017 and again in 2019, the MOHS communicated clearly with the public through regular press conferences alongside traditional and social media — something it failed to do effectively in previous health crises where the response was far less open and under-resourced [50,51].

The MOHS also mobilized in response to the 2014 Ebola and 2016 Zika outbreaks. The 2014 West African Ebola outbreak and subsequent PHEIC declaration led to Myanmar installing infrared thermal scanners in main international airports, with the CDC Myanmar supporting implementation as part of its work supporting the airport health administration. Several hospitals in Yangon were designated to receive suspected cases including Yangon General Hospital and the Waibargi Infectious Diseases Hospital. Similar airport checks and responses have been conducted for other infectious diseases such as the Middle East Respiratory Syndrome in 2015, Zika in 2016, and COVID-19 starting in early January 2020, ahead of many regional countries [52]. In mid-2019, the MOHS embarked on a five-year plan, with assistance from France, to upgrade its infectious disease diagnosis capacity, build a new bio safety level 3 (BSL-3) laboratory, and upgrade two existing facilities (i.e., tuberculosis reference laboratories) in the cities of Yangon and Mandalay [53].

The role of vertical programs and public health emergency operations centers

In the past, donor financing had skewed capacity toward vertical programs (e.g., maternal and child health, the national AIDS, tuberculosis, and malaria programs). However, recent recalibrations toward a more horizontal approach and prioritizing broader efforts to strengthen health systems, logistics and supply chain management, and information management have improved the MOHS' overall capacity. Despite this, the MOHS's approach to pandemic preparedness is heavily influenced by these vertical programs' experiences due to longstanding investments in and historically important roles played by vertical programs. The NMCP provides experience on multi-sectoral and public-private partnership approaches for the MOHS to explore partnership and resource mobilization pathways that go beyond traditional modalities and for the ministry to see health through a multi-stakeholder lens; while the NTP serves as a key reservoir of technical, diagnostic, and biosafety expertise and capacity crucial for pandemic response. In order to understand the development of PHEOCs, it is important to understand the context of Myanmar's disease surveillance and response systems.

Malaria

Malaria's historical prominence has strongly influenced the development of the country's health system. In 1927, the British established the Harcourt Butler Institute of Public Health, Rangoon, under which the Malaria Bureau was a major component [54]. While Myanmar has made significant gains since 2000, malaria remains a core epidemiological, technical, and policy focus. The vector-borne disease control unit, especially the NMCP, possesses one of the country's stronger and better-established surveillance and response systems, buttressed by a large array of implementing partners and augmented with administrative and field capacities, and technical resources. Capacities such as surveillance, rapid response teams, case and foci investigation, mass screenings and case reporting have easily transferable applications in diseases such as pandemic influenza.

Malaria elimination has also helped create a truly multi-sectoral approach, driven by the Global Fund for AIDS, Tuberculosis and Malaria's requirement for broad stakeholder participation in the country coordinating mechanism which evolved into a more inclusive coordination platform named Myanmar Health Sector Coordinating Committee (MHSCC). The MHSCC involves the NMCP, the Military's Directorate of Medical Services, EHOs operating under the aegis of non-state EAGs, donors, implementing partners, faith-based organizations, private sector, and civil society organizations. Furthermore, the NMCP has been at the forefront of experimenting with public-private partnership models involving non-health corporate entities seeking to leverage private sector resources, in-kind contributions, and expertize to strengthen surveillance and service delivery, integrate digital platforms, and promote sustainability beyond donor financing [55]. The MOHS is also transitioning approximately 15,000 village health volunteers under the NMCP and implementing partners toward integrated community malaria volunteers who will serve as frontline health workers not only for malaria, but also tuberculosis, HIV, dengue, filariasis, and leprosy to bridge health service delivery gaps [56].

Tuberculosis

Tuberculosis (TB) remains a pressing public health challenge in Myanmar, more so with the spread of multidrug resistance. Myanmar is among 30 high burden countries with prevalence twice the regional and thrice the global average [57]. The NTP — in its response to the tuberculosis epidemic and the emerging challenges of multi- and extensively drug-resistant TB cases — has developed some of the country's key preparedness proficiencies, especially for diagnostics and laboratory expertize. The two tuberculosis reference laboratories (one in Yangon and one in Mandalay) remain Myanmar's leading BSL-3 laboratories, while state/region and district diagnostics centers that house rapid TB diagnostic testing (i.e., GeneXpert) are drivers in expanding diagnostic capacity in smaller cities and towns.

The TB resources serve as the country's main reservoir of technical capacity that also serve as core components for broader diagnostics and surveillance, including specimen transport mechanisms, expertize on good laboratory practices, microbiological techniques and biosafety. Contact tracing, community-based TB care, multi-drug resistant TB response centers, and the implementation of directly observed treatment protocols have also established a human resource pool that can be utilized during the response to other public health threats. Acknowledging the numerous public health issues in Myanmar and the country's resource limitations, the NTP's accumulation of expertize and capacity are likely to frame the responses to influenza and other respiratory pandemics. Indeed, this was observed during the COVID-19 response when the MOHS leveraged tuberculosis reference laboratories and diagnostic centers to expand PCR testing capacity.

The experiences of these infectious disease programs and the accompanying funding have provided the basis for improved capacity in detection, surveillance, and relevant health infrastructure, such as laboratories and a vast network of suitably trained healthcare workers. Moreover, the requirements of much of this financing have supported the adherence to international norms, multi-stakeholder involvement and cross-border cooperation on health security issues. Together these experiences have supported the capacity development of surveillance and response for future integration with PHEOCs to operate on the frontline. While these two represent dedicated vertical programs, Myanmar's experiences in controlling infectious diseases such as trachoma, leprosy, and polio provide similar reservoirs to tap into toward strengthening the PHEOC platform.

Public health emergency operations centers

Due to seasonal influenza, H1N1, H3N1, dengue and Chikungunya outbreaks in 2019, the MOHS fast-tracked the establishment of PHEOCs under its National All-Hazard Public Health Emergency Response Plan in order to consolidate the coordination and response mechanisms under a single platform. Along with a central PHEOC headquartered in Naypyitaw, each state and region's public health department houses their own PHEOC. The development of PHEOCs has been done with support from the World Health Organization (WHO), which has also encouraged the establishment of

PHEOCs as a part of a broader trend of formalizing existing MOHS platforms for coordination, reporting, data management, and communication.

A central rapid response coordination platform currently exists for PHEOCs, chaired by the Union Minister who also leads and coordinates the state/region rapid response teams. This platform also contains state/regional public health directors, respective state/region line ministers, and a technical working group. Township Medical Officers serve as the central focal points for the coordination platform at the township level — Myanmar's smallest and main self-contained administrative unit — and oversee the township's health system and department from the township center down to the field volunteers in surrounding villages. Due to staff shortages, the township medical officers often assume responsibility for both the medical and public health aspects of a response. Information sharing is done via groups established on the messaging platform Viber.

Further strengthening PHEOCs will establish a dedicated platform to improve the MOHS' inter-departmental collaboration, emergency preparedness, and response capacity for public health emergencies, such as pandemics. This will enable the formation of a cross-cutting approach with dedicated staff for more efficient inter-sectoral coordination and response, rather than relying on ad-hoc groups. It will also allow for PHEOCs to be used for other activities in "peace time" and to address other key public health challenges that necessitate strong leadership, interventions and responses using continuous data monitoring, and analysis of highly fluid situations.

The PHEOCs during COVID-19 response through to July 2020

Myanmar had its first confirmed COVID-19 case on March 23, 2020 though surveillance measures had already been in place since early January and the MOHS presented a proposal to mobilize resources including activities for enhanced surveillance in early February. The PHEOCs were mobilized during the response to COVID-19 and undertook a variety of functions including holding regular digital meetings, implementing biosafety protocols, and coordinating community surveillance, community quarantine, and health education activities.

The role of the nascent PHEOCs changed considerably in three ways in the first half of 2020 amid the pandemic. First, the 2019 discussions on PHEOCs at the national and state/regional levels supported the initial public health response to COVID-19 while the established physical infrastructure was engaged to coordinate initial responses, such as communications and training.

Second, the PHEOC platform, situated within the MOHS, was superseded by a whole-of-government response, that included the military and led by the National-Level Central Committee for Prevention, Control and Treatment of Coronavirus Disease 2019. For all intents and purposes, this served as a centrally organized PHEOC. State/regional-level committees were also formed, and in some instances village level committees, though the latter have now been mostly dissolved.

Third was the expansion and upgrading public health departments in key regional cities to address the both the diagnosis and management aspects of the pandemic. Cities such as Mawlamyine, Taunggyi, Monwya were tasked with the management of sub-national COVID-19 responses. As part of this, they established or expanded diagnostic laboratories for COVID-19 and mobilized all medical staff toward COVID-19-related activities, such as training field volunteers, overseeing quarantine facilities, and ensuring the continuation of public health programs [58]. While some of the upgraded health departments continued to operate on their original catchment areas, others served as nodes for larger areas, or for specific border crossing points (e.g., Mawlamyine served as the main control node not only for Mon State but also for Kayin State, and the Myawaddy-Mae Sot border crossing).

Complementing this, legislation to improve legal frameworks for combatting diseases were amended. In May of 2020, a revised version of the 1995 Prevention and Control of Communicable Diseases Law was submitted to parliament. The MOHS was also conferred sweeping powers, in coordination with the General Administration Department that oversees local governance, for surveillance and implementing localized lockdowns. The PHEOCs also served in outreach efforts, especially in assisting the MOHS develop health education material in local languages for circulation on social media platforms, such as Facebook.

Conclusions: strengthening Myanmar's urban pandemic preparedness and lessons for other low and middle-income countries

Implementing a dedicated PHEOCs platform will be a significant improvement to Myanmar's response to urban and nationwide public health emergencies. Experiences and lessons learned from operationalization of the PHEOCs will augment the overall pandemic response and strengthen capacity. As already noted in this chapter, there are many areas that warrant improvement in cities in Myanmar to better prepare for major public health emergencies. These include health system strengthening, specific improvement to laboratory and diagnostic facilities, human resource capacity building, and public-private partnerships.

Health system strengthening

A key issue is the need to transition from viewing PHEOCs solely as physical infrastructure toward regarding them as a broader package encompassing technical capacity, inter-departmental coordination, public relations and crisis communications management, and the physical as well as soft skills infrastructure needed for smooth and effective operation in times of crisis [59]. In order to be able to respond to an epidemic effectively, a comprehensive PHEOC system should be prepared for: (1) sentinel surveillance system with real-time notification, including established communications with primary care centers and medical facilities from both the public and private healthcare sectors; (2) screening, detection and referral of suspected cases at various potential high-risk 'emergence points' such as airports, border checkpoints and high-risk urban and peri-urban locations; (3) diagnosis and case management with isolation facilities, and management of such facilities; (4) intensive care units and related infectious control or biosafety protocols; (5) health supply chain management; and (6) risk communication, inter-ministry coordination and community engagement.

Each component, arranged in all instances involving infectious pathogens, would be implemented according to guidelines from the MOHS and the WHO. It is imperative to protect healthcare staff, patients, law enforcement, and visitors from exposure with standard precaution procedures that may require periodic reassessment. Such measures will be not only for case containment but also to preserve "frontline" health worker morale and mitigate tensions between policy leadership and healthcare workers [60]. Community system strengthening is also important to build trust between the MOHS and local communities, and bolster preparedness at the local-level through awareness raising, self-quarantine of people exposed to priority pathogens, risk reduction measures (e.g., respiratory etiquette, hand washing and social distancing), to prevent and address social stigma and panic associated with mis- and disinformation [61,62].[g]

Strengthening laboratory and diagnostic facilities

The primary focuses of laboratory strengthening in Myanmar's urban areas should be to ensure personnel safety and timely quality diagnostics in their jurisdictions. Each laboratory and diagnostic facility will have to perform local risk assessments for each process steps such as sample collection, transport and reception, diagnostic procedures (e.g., PCR), and waste disposal. Myanmar's efforts to expand and establish new BSL-3 laboratories will require not only technical upskilling, but also more stringent protective equipment, biosafety protocols, and management capacities. Strengthening can include refreshers' training of laboratory personnel according to guidelines from the WHO and national authorities, findings from risk assessment surveys, a dedicated biosafety inspection, and quality assurance regime.

Surge capacities — such as drawing on capacity of the MOHS's Department of Medical Research, academic departments, and private and military-run

---

g. In March 2020, voice messages purporting cases — before the first confirmed case was announced, and also afterward — triggered panic buying across major cities.

diagnostic laboratories — will also need to be identified and planned in advance. Facilities should have adequate stocks of appropriate disinfectant and decontamination materials along with appropriate hazard and leakage response mechanisms for eventualities such as fires, earthquakes, and accidents. This was evident in the rapid upscaling of the National Health Laboratory in Yangon and the Public Health Laboratory in Mandalay in response to COVID-19 that included acquisition of new PCR testing machines. This complements a five-year European Union and French government supported renovation of the National Health Laboratory announced in September of 2019.

Human resource capacity building for emergency response

Frontline health workers inherently deal with the first few cases of any public health emergency before an outbreak has been identified. These workers may not be well-prepared for initial infection control, and risk becoming super-spreaders to other patients and health workers if they are not properly trained. Health human resource capacity building requires more than just training on the diagnosis and management of infectious diseases but should also focus on practices that prevent transmission within the healthcare setting. While in Myanmar these are human resource challenges are uniquely felt in rural areas, improvements are equally needed in urban areas to improve health outcomes.

These should be complemented with the public health aspect related to infection control, such as health education campaigns, referral systems, and community-based management. Another issue to consider is health workers' welfare management during emergencies to improve the effective deployment of health workers, reduce infection exposure risks and work stress, and maintain general well-being.

The lack of proper training in managing epidemic and pandemic settings can weaken or even collapse fragile health systems where the limited resources will be quickly consumed during emergencies. And if frontline health workers become infected due to inadequate protection, the health system could face further shortage of health workers and a corresponding collapse in health worker morale that can cascade into more deleterious eventualities.

In Myanmar, civil society played a crucial role in the community response to the COVID-19 pandemic by managing localized lockdowns, overseeing quarantine facilities, providing community ambulances, transporting suspected COVID-19 patients to designated facilities, and providing free funeral services. The MOHS and health-focused NGOs provided emergency training for these specialized CSOs in order to promote coordination and reduce transmission risk.

Public-private partnerships

Public-private partnerships (PPP) seek to leverage the resources, networks, and expertize of the private sector through in-kind contributions and partnerships.

Such partnerships offer a pathway for multi-sectoral and effective pandemic preparedness in low resource settings and are a valuable tool for urban pandemic preparedness.

In urban settings, PPPs can be used to bolster the MOHS's service delivery, production of emergency equipment, sourcing and procurement of medical commodities, and transport of essential supplies. PPPs with private clinics, hospitals and medical suppliers can also assist in the provision of surge capacities to respond to public health emergencies. This will be crucial in Myanmar's urban centers that still confront inadequate healthcare infrastructure despite recent investments [63]. For example, in response to the COVID-19 pandemic's arrival in Myanmar, non-health businesses have stepped in to procure, pool, and donate urgently required personal protective equipment, breweries have shifted to the production of hand sanitizers, and religious institutions have offered space for quarantine. The MOHS also engaged in a PPP to establish a dedicated COVID-19 medical center in Yangon's outskirts that involved the Construction Entrepreneurs Association, the Myanmar Private Hospitals' Association, the Pharmaceutical and Medical Equipment Entrepreneurs Association alongside the Red Cross, civil society, and donors [64].

As such PHEOCs will need to develop multi-stakeholder coordination capacities with non-governmental organizations and the private sector along with identifying resources and needs, and to develop an operational culture that embraces inter-operability with the private sector that may approach issues from very different perspectives.

Lessons from Myanmar for other LMICs

Myanmar's geography, connectivity to other LMIC countries, and zoonoses hotspots render improving pandemic preparedness a key priority. The progress made to date in improving capacities needs to be further supported both domestically and by international organizations. In particular, the country's experience in leveraging existing capacities established by vertical programs may be informative for other pandemic preparedness efforts in urban environments and could serve as a viable model for cities in resource-limited settings.

References

[1] Nuclear Threat Initiative, Johns Hopkins Center for Health Security. The Economist Intelligence Unit. Global Health Security Index: 2019 GHS index country profile for Myanmar. 2019. www.ghsindex.org/country/myanmar/. [Accessed 30 March 2020].

[2] Madhav N, Oppenheim B, Gallivan M, Mulembakani P, Rubin E, Wolfe N. Pandemics: risks, impacts, and mitigation. In: Jamison DT, Gelband H, Horton S, Jha P, et al., editors. Disease control priorities: improving health and reducing poverty. 3rd ed. Washington: World Bank; 2017.

[3] Karmacharya B. Myanmar can learn from its neighbours in managing rapid urbanization. Myanmar Times; October 01, 2018.

[4] Mirante E. A plague of epidemics throughout Myanmar history. New Mandala; April 01, 2020.

[5] Luce GH. Old Burma-early pagan, vols. 1–3. Locust Valley: J.J. Augustin; 1969.

[6] Ko K, Lwin K, Thaung U. Conquest of scourges in Myanmar. Yangon: Myanmar Academy of Medical Science; 2002.

[6a] Razavi A, Erondu NA, Okereke E. The Global Health Security Index: what value does it add? BMJ Glob Health 2020;5:e002477.

[7] The World Bank. Ebola response fact sheet. 2016. http://www.worldbank.org/en/topic/health/brief/world-bank-group-ebola-fact-sheet. [Accessed 30 March 2020].

[8] Keogh-Brown MR, Smith RD. The economic impact of SARS: how does the reality match the predictions? Health Pol 2008;88:110–20.

[9] Asian Development Bank. ADB briefs: an updated assessment of the economic impact of COVID-19, No. 133. Manila: Asian Development Bank; 2020.

[10] Eckstein D, Hutfils M-L, Winges M. Global climate risk index 2019: who suffers most from extreme weather events? Weather-related loss events in 2017 and 1998 to 2017. Bonn: Germanwatch; 2019.

[11] United Nations Environment Programme. Learning from Cyclone Nargis: investing in the environment for livelihoods and disaster risk reduction. Nairobi: UNEP; 2009.

[12] Food and Agriculture Organization. New Myanmar one health (OH) strategy. FAO; March 17, 2016.

[13] Bordier M, Roger F. Zoonoses in South-East Asia: a regional burden, a global threat. Anim Health Res Rev 2013;14:40–67.

[14] United Nations Office for Coordination of Humanitarian Affairs (OCHA). Interagency Emergency Response Preparedness (ERP) plan for Myanmar. New York: UNOCHA; 2017.

[15] Ministry of Health and Sports. National strategic plan for prevention and control of avian influenza and human influenza pandemic preparedness and response. Naypyitaw: MOHS; 2006.

[16] Ministry of Health and Sports. Early Warning, Alert and Response System (EWARS): standard operating procedure. Naypyitaw: MOHS; 2018.

[17] Bannister-Tyrrell M, Viney K, Happold J, Zalcman E, Alene K, Wangdi K, et al. State of the region 2019: health security in the Indo-Pacific. Canberra: Indo-Pacific Centre for health security; 2019. Australian Department of Foreign Affairs and Trade, Centre for Health Security. 2019. https://indopacifichealthsecurity.dfat.gov.au/state-region-2019-health-security-indo-pacific. [Accessed 30 October 2020].

[18] Phua J, Faruq MO, Kulkarni AP, Redjeki IS, Detleuxay K, Mendsaikhan N, et al. Critical care bed capacity in Asian countries and regions. Crit Care Med 2020;48:654–62.

[19] Aung T, Lwin MM, Sudhinaraset M, et al. Rural and urban disparities in health-seeking for fever in Myanmar: findings from a probability-based household survey. Malar J 2016;15:386.

[20] The Asia Foundation. 2020 union expenditure by functional classification – Myanmar budget dashboard. 2020. https://mmbudgets.info/#/2020/Union/sectors>. [Accessed 30 March 2020].

[21] Soe APK. Ministry details plan to address important health issues in 2020. Myanmar Times; January 03, 2020.

[22] Ergo A. Resource mobilization options for Myanmar. Naypyitaw: Knowledge Sharing Session on Health Financing; 2018.

[23] Carroll J. Yangon healers peddle 'miracle' cancer cure. Frontier; March 01, 2016.

[24] Htoo TS. Overview of health financing situation in Myanmar. Naypyitaw: Knowledge Sharing Session on Health Financing; 2018.

[25] Sein TT, Myint P, Cassels A. Policy note 2: how can health equity be improved in Myanmar? Manila: WHO Regional Office for the Western Pacific; 2015.

[26] Sein TT, Myint P, Cassels A. Policy Note 3: how can the township health system be strengthened in Myanmar? Manila: WHO Regional Office for the Western Pacific; 2015.

[27] The World Bank. Myanmar's urbanization (Vol. 3): creating opportunities for all - full report. Washington: World Bank Group; 2019.

[28] The World Bank. Population in urban agglomerations of more than 1 million (% of total population) − Myanmar. 2018. https://data.worldbank.org/indicator/EN.URB.MCTY.TL. ZS?locations=MM. [Accessed 30 March 2020].

[29] The World Bank. Myanmar living conditions survey 2017 − report 03: poverty report. Washington: World Bank Group; 2019.

[30] Matelski M, Sabrié M. Challenges and resilience in Myanmar's urbanization: a special issue on Yangon. Moussons 2019;33:11−31.

[31] The Economist. Myanmar's countryside is emptying and its cities bursting. The Economist; July 11, 2019.

[32] Htin K. Yangon's squatter population remains high. The Nation; October 14, 2015.

[33] Thar H. Government seeks to tame Hlaing Tharyar, Yangon's wild west. Frontier; November 06, 2019.

[34] Khaing TT. Urbanization: the structures of sustainable urban landscape of Myanmar. Chiang Mai: Chiang Mai University; 2015.

[35] Htoon KL. Going underground: upgrading Yangon's colonial-era sewers. Frontier; May 29, 2018.

[36] Heang Chhor H, Dobbs R, Hansen DN, Thompson F, Shah N, Streiff L. Myanmar's moment: unique opportunities, major challenges. New York: McKinsey Global Institute; 2013.

[37] The Asia Foundation. Insight into urban well-being in Myanmar: the 2018 city life survey. San Francisco: The Asia Foundation; 2018.

[38] Lwin N. Myanmar told to brace for major COVID-19 outbreak as returning migrant among new cases. Irrawaddy; March 30, 2020.

[39] Lynn HK. Digging deeper: inside the Hpakant jade mine. Myanmar Times; June 08, 2016.

[40] Oh SA. Internal migration in Myanmar: patterns, benefits and challenges. Singapore: ISEAS-Yusof Ishak Institute; 2019.

[41] Carrara VI, Lwin KM, Phyo AP, Ashley E, Wiladphaingern J, Sriprawat K, et al. Malaria burden and artemisinin resistance in the mobile and migrant population on the Thai-Myanmar border, 1999-2011: an observational study. PLoS Med 2013;10:e1001398.

[42] Health Information System Working Group. The long road to recovery: ethnic and community-based health organizations leading the way to better health in Eastern Burma. Naypyitaw: Health Information System Working Group; 2015.

[43] Hellmann N. Mongla: 'the wildlife trafficking capital of the world'. Irrawaddy; October 13, 2016.

[44] Myint YY. MP seeks relocation of Mingalar Taung Nyunt poultry market. Myanmar Times; March 12, 2018.

[45] Zaw T. Visions of hell: inside Yangon's pig slaughterhouses. Frontier; November 3, 2019.

[46] World Health Organization. Joint external evaluation of IHR core capacities of the Lao people's democratic republic. Geneva: WHO; 2017.

[47] World Health Organization. Joint external evaluation of IHR core capacities of the Republic of the Union of Myanmar. Geneva: WHO; 2017.

[48] Abimbola S, Baatiema L, Bigdeli M. The impacts of decentralisation on health system equity, efficiency, and resilience: a realist synthesis of the evidence. Health Pol Plann 2019;34:605−17.

[49] Brennan E, Abimbola S. Understanding and progressing health system decentralisation in Myanmar. Global Secur Health Sci Pol 2020;5:17−27.

[50] Lone W, Lee Y. Myanmar H1N1 swine flu death toll rises to 10. Reuters; July 31, 2017.

[51] Soe APK. H1N1 flu claims 108 lives in first eight months. Myanmar Times; September 04, 2019.

[52] Soe APK. Myanmar intensifies screening of travellers over China pneumonia outbreak. Myanmar Times; January 9, 2020.

[53] Soe APK. Government to upgrade national health lab. Myanmar Times; September 30, 2019.

[54] Harcourt Butler Institute of Public Health. Second annual report of the Harcourt Butler Institute of Public Health, Rangoon for the year 1927. Rangoon: Government Printing and Stationary; 1928.

[55] Lwin S, Minn K, Wai KS. Myanmar's multisectoral approach to malaria elimination. Manila: Asia Pacific Leaders Malaria Alliance; 2019.

[56] United Nations Office for Project Services. RAI2E empowering communities through ICMVs. 2019. https://raifund.org/en/news/rai2e-empowering-communities-through-icmvs. [Accessed 30 March 2020].

[57] National Tuberculosis Program. Annual report, 2016. Naypyitaw: MOHS; 2018.

[58] Thar H. Swabs, staff and supply chains: how Myanmar cleared resource hurdles to ramp up COVID-19 testing. Frontier; June 10, 2020.

[59] World Health Organization. Handbook for developing a public health emergency operations centre − part A: policy, plans and procedures. Geneva: WHO; 2018.

[60] Chersich MF, Gray G, Fairlie L, Eichbaum Q, Mayhew S, Allwood B, et al. COVID-19 in Africa: care and protection for frontline healthcare workers. Glob Health 2020;16:46.

[61] Jaiswal J, LoSchiavo C, Perlman DC. Disinformation, misinformation and inequality-driven mistrust in the time of COVID-19: lessons unlearned from AIDS denialism. AIDS Behav 2020:1−5.

[62] Htun Z, Ei KK, Gerin R. Rampant fake news is a key front in Myanmar's coronavirus battle. Radio Free Asia; March 27, 2020.

[63] World Health Organization. Global health observatory data repository. Geneva: WHO; 2017.

[64] Soe APK. Yangon opens new facility to treat up to 2000 COVID-19 patients. Myanmar Times; April 22, 2020.

Chapter 9

Preparedness through wargaming: Urban Outbreak 2019 and its applicability to America's response to the COVID-19 pandemic

Hank J. Brightman[1]
College of Maritime Operational Warfare/Civilian-Military Humanitarian Response Program,
U.S. Naval War College, Newport, RI, United States

One of the greatest challenges for personnel working within the humanitarian space is adequately preparing for a response during multifaceted operations. No environment portends a greater need for informative, deliberate data to be available to senior decision-makers than in response to biological threats in an urban environment. Indeed, while factors such as the physical environment, population density, intrinsic culture, and socioeconomic status all present singular trials, there are few venues where these complicated dynamics are more intensely and collectively tested than in infectious disease response within a megacity (i.e., a city with a population of more than 10 million people). This chapter focuses on the development of a wargame simulating an outbreak in a fictitious megacity, discusses the key insights and findings of the simulation, and summarizes how the lessons learned may apply to the response to COVID-19 in cities.

A spate of academic studies and journal articles exists devoted to pathogen spread in urban environments including megacities in the developing world. Traditional scholarship is an integral component of understanding myriad catalysts to instability present on the urban landscape. This project examined many of these studies as part of the selective coding process within the

---

[1] Please note that the views and opinions expressed in this chapter are purely those of the author, and do not represent official policy or positions of the Naval War College, Department of the Navy, or the United States Government.

Inoculating Cities. https://doi.org/10.1016/B978-0-12-820204-3.00009-7

grounded induction method employed in this research effort. However, to better understand the complexity contagion presents within megacities, the use of wargaming provides practitioners, researchers, and leaders tasked with responding infectious disease outbreaks in populous urban environments with a shared phenomenological event more appropriate for robust decision-making. The data obtained by playing through these notional scenarios — leveraging experts in their respective fields —affords those engaged in real-world response with a palette of insights that may prove informative to making rapid decisions in-crisis.

It is important to note that such scenario-based conversations, referred to throughout this chapter as wargames, because this approach was borne through associated studies in military science in the mid-to-late nineteenth century, are not designed to be predictive [1]. Wargames are as much an art as a science, and their overarching utility rests in an ability to bring players together within a common, "safe container" where they "may develop strategies and contingencies" [1]. Moreover, wargames are not experiments, since their underpinnings draw heavily upon the perspectives of subject matter experts. With respect to the response to a biological pathogen event in a megacity, given the requisite knowledge base, players are not randomly selected from a population that is known to be normally distributed, which means there is little, if any, generalizability or inferentiality to the findings of such inquiry [1]. Although not predictive, the individual and collective insights of participants in such games bear tremendous fruit for senior decision-makers engaged in activities such as humanitarian response to an infectious disease in a megacity.

Origins of the Urban Outbreak 2019 Wargame

Wargames provide a viable approach to focus the interests of subject matter experts on complex issues when no clear solution is evident. Accordingly, it is unsurprising that following more than two years of discussion among the spate of participants in the annual pandemics and urban issues working groups, held as part of the 2017 and 2018 Civilian-Military Humanitarian Response workshops hosted at Brown University in Providence, Rhode Island, wargaming would emerge as the preferred methodology to explore the issues and challenges encountered in a notional response to an infectious outbreak in a fictitious megacity of a developing nation. Both working groups are linked to the partnerships between the United States Naval War College's (NWC) Civilian-Military Humanitarian Response Program (HRP), Brown University's Center for Human Rights and Humanitarian Studies, and Harvard Humanitarian Initiative, each of whom has a key stakeholder interest in promulgating applied research to inform senior decision-makers. These periodic humanitarian response workshops serve as a catalyst to move from conversation to action, as evidenced by the genesis of the Urban Outbreak

2019 Wargame, which was fostered in the discussions of the 2017 and 2018 working groups ultimately became operationalized in a two-day event at Johns Hopkins University (JHU) Applied Physical Laboratory (APL) in September 2019.

Execution of a wargame is a complex process involving an exhaustive series of tasks, many of which must be carefully synchronized to ensure a successful project outcome. For example, prior to the development of game design and scenario development, great care must be placed on the specific problem to be addressed, along with the game's overarching purpose and objectives. Research questions and specific areas of inquiry must be inextricably linked to the wargame's objectives and design, and data capture mechanisms must be developed that seamlessly integrate with the game design to keep players grounded in the wargaming scenario and their decision-making processes [2]. In addition, bringing the right players to the game to feed this data loop is essential, which includes attending to the physical needs of the players before, during, and after game play. Accordingly, the role performed by logistics personnel in attending to these requirements is also a critical facet of a successful wargame.

To bring the Urban Outbreak 2019 wargame to fruition, a triumvirate was formed between the NWC's HRP, Uniformed Services University of the Health Sciences (USUHS) - National Center for Disaster Medicine and Public Health (NCDMPH) and the JHU APL. The NWC-HRP played a vital role in designing the wargaming and providing its analytic framing, while USUHS-NCDMPH identified essential medical and public health personnel who would be engaged in response to a notional infectious outbreak in a megacity. The NCDMPH also provided funding for many of these participants to travel to the game. About 50 participants representing public health, emergency medicine, supply and logistics, and healthcare interests were identified as players, along with government representatives (including US and international militaries), and non-governmental/international organizations (NGOs/IGOs).

Lastly, the APL hosted the two-day event at its facility in Laurel, Maryland, and supported the development of a notional infectious pathogen based on a modified, treatable pneumonic plague, created effective visualizations of the fictitious affected state and city (referred to as Olympus and Olympia, respectively), and built the player surveys and other data collection instruments required to support post-game analysis of common themes and insights.

Areas of research inquiry and scenario considerations

Fully embracing the concerns of both the pandemics and working group and urban issues working groups stemming from the 2017 and 2018 Civilian-Military Humanitarian workshops hosted at Brown University, the wargaming team focused on the following three broad objectives (Box 9.1).

BOX 9.1 Primary objectives of the Urban Outbreak 2019 Wargame

(1) Identifying priorities for organizations involved in such a response as well as their associated strengths and weaknesses;
(2) Identifying civilian and military agency coordination and communication challenges and opportunities for preparedness and response;
(3) Developing a set of questions, derived from player activities and observations made during the exercise, to inform a research and training agenda [3].

Aligned with these objectives, the design team developed a robust scenario emphasizing three specific periods of response to an infectious pathogen in a densely populated urban environment. These periods (i.e., moves) considered the outbreak and initial response (move one), pathogen apex and associated cascading failures (move two), and outbreak decline and re-deployment of military forces to their regular operations (move three). A highly infectious notional pathogen which was loosely based on pneumonic plague was introduced in the scenario. Although readily treatable, success in patient recovery was closely aligned with the availability of public health infrastructure and rapid deployment of medical personnel, allowing the game design to replicate the challenges posed by the lack of such resources in a notional megacity within an affected state typified by a developing nation. Moreover, the pathogen and its associated transmission rate were deemed viable as a Public Health Emergency of International Concern requiring a "prompt, robust international humanitarian response" [3].

For the fictitious city of Olympia within the fabricated nation of Olympus, the game design team sought to create a plausible environment where a lack of resilient public health infrastructure and redundant medical capability would allow game players to examine cascading failures across multiple humanitarian sectors. Beyond the physical layout of this city, which is comprised of more than 21 million inhabitants (60% of whom live in slum-like conditions) there is vast economic disparity between groups within the megacity. Only three percent of Olympia's residents have access to public water facilities. Further complicating life within this notional urban environment is the reliance on informal (i.e., unlicensed) medical care and environmental contamination including illicit disposal of hazardous and medical waste [3].

From a law and order perspective, even within the wealthiest environs of Olympia's secure, gated communities, there is burgeoning personal crime (including kidnapping perpetrated by roving violent gangs), institutional corruption, and disparate enforcement of protection of personal property; the latter of which is usually reserved for the wealthiest citizens or those with a significant connection to public officials or state-supported financial enterprises. Many of the slum-like areas of Olympia, as well as the rural portions of

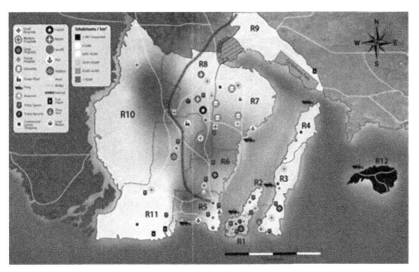

FIG. 9.1 Visualization of affected state of Olympus including megacity of Olympia in Sectors R5-R6 [3]. Source: Urban Outbreak 2019 Game Scenario Brief (U.S. Government — Naval War College). https://digital-commons.usnwc.edu/cgi/viewcontent.cgi?article=1000&context=civmil response-program-sims-uo-2019.

Olympus, are protected by warlord-like organized crime groups aligned with familial blood ties, ethnocentric homogeneity, or sectarian beliefs. A broad visualization of the affected state (Olympus) and the massive population density of its megacity, Olympia is provided in Fig. 9.1.

It is within the context that the infectious pathogen depicted in Fig. 9.2 has found its way into the nation of Olympus and its megacity, Olympia. Rapidly spiking infection rates with greater than 45,000 ill within days of the initial outbreak, and mortality rates for untreated victims occurring within six days of contact, the outbreak presents at the onset with "fever, cough, headache, fatigue, and general malaise; progressively worsening to include bloody sputum production, respiratory distress syndrome, respiratory failure, and death" [3]. Although readily treatable with injectable antibiotics, the nation of Olympus and megacity, Olympia possess neither sufficient stores of these medicines, nor a viable public health delivery system to ensure effective mitigation or containment of the disease.

Game design and player move characteristics

The Urban Outbreak 2019 Wargame sought to bring together key stakeholders from governmental sectors, non-governmental/intergovernmental organizations, and the private sector, to explore a notional, complex response to a fictional infectious disease within the urban center of a developing nation. The

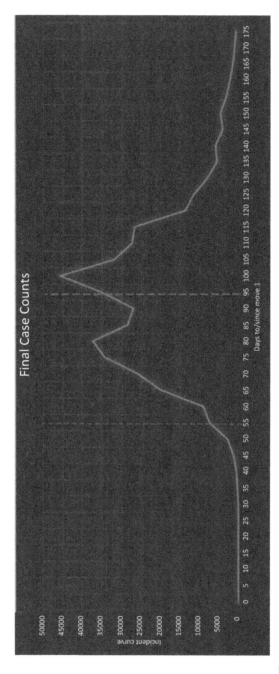

FIG. 9.2 Pathogen spread of notional disease in Urban Outbreak 2019 Wargame [3]. Source: Urban Outbreak 2019 Game Scenario Brief (U.S. Government – Naval War College). https://digital-commons.usnwc.edu/cgi/viewcontent.cgi?article=1000&context=civmilresponse-program-sims-uo-2019.

overarching emphasis of the game focused on mission prioritization and the coordination of efforts between medical and public health professionals, government entities, militaries, and private sector actors, while considering their proprietary, diverse perspectives regarding an intricate problem set. The game was comprised of approximately 50 experts from each of these sectors, including a cadre of military personnel from four nations selected for their experience in security, planning, and logistics. Player activities emphasized the challenges and opportunities borne working toward the unity of effort with a broad range of interests in a hyper-dynamic, risk-infused environment. During two full days of game play, participants aspired to ensure that the needs of their respective sector were met while engaged in initial response (move one), outbreak escalation (move two), and post-apex trending toward normalization with the redeployment of foreign militaries out of the affected state (move three).

At the start of game play, participants were assigned to one of five initial cells (i.e., medical, public health, government, militaries, and private sector) based on their sector's respective characteristics. Once seated in the sector-based gaming spaces, players were invited to identify a maximum of three priority activities and up to five essential stakeholders (i.e., external to their organization) necessary to complete this activity [3]. In addition to collectively identifying these three activities and five stakeholders using a group consensus process, each player also completed an individual survey focused on their organization's priority activities, essential stakeholders, and conducted a self-assessment of their cell's response efficacy at this point in the game.

Following the first move, all five cells rejoined in a common plenary session where their efforts were self-adjudicated collectively. Players were then afforded the opportunity to return to their previously assigned sector cell; or, if desired, they were given the freedom to switch to a different sector cell that better aligned with their perceived priority activities and essential stake-holders at this point in the response. The entire process was repeated one additional time on the second day of game play, with players coming together in a final group move and holistic plenary session immediately following move three. A summary of those stakeholders deemed significant during each of the first two phases of the response operation is provided below.

As visualized in Fig. 9.3, it is noteworthy that at the onset of game play, participants strongly connected with the local Ministry of Health in Olympia to serve as a priority stakeholder in the response to the outbreak within the megacity. Interestingly, national assets were deemed far less essential, although game participants identified international assets (e.g., the World Health Organization and multinational forces from the United States, United Kingdom, Germany, and Chile) as essential to initial response operations. Moreover, as depicted in Fig. 9.4, when the intensity of the outbreak increased in terms of both scope and scale, the call for the multinational force became

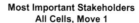

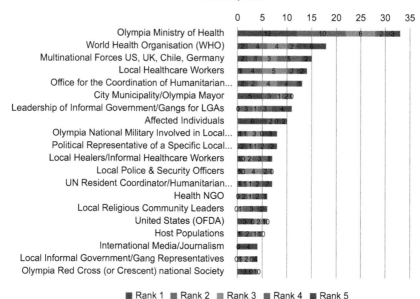

FIG. 9.3 Player-identified most important stakeholders in move one of Urban Outbreak 2019 Wargame [4]. Urban Outbreak 2019 Pandemic Response: Select Research & Game Findings April 2020: Newport, RI: Defense Automation and Production Service. https://digital-commons.usnwc. edu/cgi/viewcontent.cgi?article=1001&context=civmilresponse-program-sims-uo-2019.

even greater, with military assets deemed the essential stakeholder along with the United Nations Office for the Coordination of Humanitarian Affairs (OCHA).

Although the game's design structure precluded assessment of what assets would receive prioritization in the pathogen's post-peak decrease and the nation's return toward stabilization, comments garnered from player reportage suggest that efforts toward greater infrastructure development (i.e., both physical infrastructure and public health systems) were unlikely. In short, both Olympia and Olympus post-outbreak would likely appear much like the affected state and megacity had looked before the initial outbreak had occurred. Such findings are important because they speak to the real-world challenges encountered in epidemic and pandemic response. Beyond the handful of NGOs and private interests expected to remain in Olympus and Olympia at the close of move three, very little effort was made to address future outbreaks proactively. Given that the 50 players who participated in this

**Most Important Stakeholders
Move 2, All Cells**

FIG. 9.4 Player-identified most important stakeholders in move two of Urban Outbreak 2019 Wargame [4]. Source: Urban Outbreak 2019 Pandemic Response: Select Research & Game Findings April 2020: Newport, RI: Defense Automation and Production Service. https://digital-commons.usnwc.edu/cgi/viewcontent.cgi?article=1001&context=civmilresponse-program-sims-uo-2019.

game averaged 10 years of experience in real-world humanitarian response operations, these insights are significant when considering the cost-benefit of engaging in international humanitarian operations at a strategic level [4].

Key insights and findings

The overarching purpose of this wargame was to explore issues and challenges identified in the response to an infectious outbreak in a notional megacity within a developing nation. In scoping this project, all three sponsors (i.e., NWC-HRP, USUHS-NCDMPH, and APL) expressed their unwavering support for bringing NGOs and IGOs into this event. Such entities are rarely represented in wargames involving military and US governmental interests due to their need to strictly conform to the four core humanitarian principles of humanity, neutrality, impartiality, and independence. To respect these principles, the wargame employed a strict non-attribution policy, making it clear that it was the collective insights of the 50 participants that would drive future research and training.

Findings were garnered using the qualitative analytic methods identified in the game's Data Collection and Analysis Plan. Specifically, the Urban Outbreak 2019 Wargame employed an inductive, qualitative research design, applying a mixed-methods approach. A wealth of data were collected using both externally and internally validated survey instruments designed to capture player-derived insights. A common method of reportage was also used to capture cell-based and group plenary discussions [5]. Game data included approximately 168 pages of cell and plenary-based reportage notes, all of which focused on discussions within the five player cells and collective plenary sessions. In addition, the findings of nine cell-based surveys, two post-move visualizations were included in the datasets for post-game analysis (Figs. 9.3 and 9.4).

In post-game analysis, text-based products were coded using a list of nine words identified during the selective coding (i.e., literature review) portion of this project. The nine words included as codes were coordination, communication, priorities, challenges, opportunities, prepare, response, strength, and weakness. Using a grounded induction analytic tool, analysts identified the strongest in-vivo co-occurrences in the second move of the game (i.e., during the height of the outbreak's spread within Olympia and surrounding environs). Specifically, players expressed their frustrations over challenges in response, namely around issues of coordination and communication. Much of the tension in communication stemmed from a lack of shared technical capabilities between NGOs/IGOs and multinational forces. They lacked the ability to speak on common communications networks such as collective radio frequencies for delivering lifesaving supplies and equipment. Even the websites used by the multinational forces to input real-time data to support logistics and sustainment differed from the more archaic spreadsheet-based system employed by military players to track assets and identify the readiness of platforms and capabilities for aid delivery. There was also an overarching lack of both coordination and activity prioritization by the OCHA, Olympia's Ministry of Health, or the national government of the affected state (Olympus). It should be noted, however, that because a broad-based command and control structure was neither assumed nor provided to players as part of this game's design, it is possible that this manifested as an artifact of game play to a greater extent than would be anticipated in a real-world response to such an outbreak, particularly if the United Nations and multinational forces were included in such operations.

Beyond the use of grounded induction to identify key player themes, five collective insights aggregated through analysis of individual surveys and reportage notes emerged in post-game research. Each of these insights is discussed in greater detail in this portion of the chapter [4].

Insight one: social behaviors matter immensely

In a complex environment such as the megacity presented in this game, diversity abounds. Ethnocentric overidentification with a religious belief system, political faction, or an appealing physical characteristic or trait is common, even when such factors may not represent most of the city's population (e.g., a small, but influential ruling minority group). However, to respond to pathogen spread efficaciously, each of these cultures and subcultures will need to be considered, and a common strategy for mitigating disease propagation must be developed. For example, masking-wearing, hand-washing, and other strategies to ensure best practices for reducing pathogen spread must be considered in the context of the beliefs and values of the members in this diverse urban landscape. Employing anthropologists, sociologists, political scientists, and theologians familiar with these cultures and subcultures is imperative for developing strategic messaging around public health. Moreover, rather than having these experts disseminate such messages themselves, they should leverage the collective voice of both public officials and informal leaders such as community elders.

Fostering a thoughtful, empathic discussion regarding how and why societal behaviors may need to change in the wake of the outbreak while ensuring sensitivity to a broad range of beliefs, values, and norms is critical to curtailing additional challenges to an effective response.

Insight two: every failure in a system impacts every other system

In response operations, such as those presented in the Urban Outbreak 2019 Wargame, each sector possesses its own inherent structures and systems. However, these systems are inextricably linked to every other system through a complex web of interdependencies. These relationships are so fragile that one slight perturbation in a single sector, such as logistics, can have ripple effects that adversely impact systems in other sectors, such as food security for a vulnerable population located outside Olympia's city center. Identifying, understanding, and visualizing these interdependencies is essential to effective planning and the delivery of aid during periods of crisis.

Insight three: the private sector is vital in response

The private sector possesses a nimbleness that makes it uniquely suited for all phases of response operations. From logistics to health care, private sector interests can deploy rapidly without the cumbersome trappings of governmental structures and political entities. While private-sector capabilities, such as private security contractors, were identified specifically as an important priority stakeholder in the second move of the game (i.e., during the apex of the outbreak), it should be noted that a substantial portion of the final plenary focused on the value of private sector entities in the post-outbreak stabilization and rebuilding phase, particularly as multinational forces, and UN/IGOs redeployed in support of their own national missions.

Great Power Competition theory posits that both China and Russia seek to expand their global influence by garnering exclusive rights to port access, military basing, telecommunications infrastructure, and agricultural, mining, and natural resource stores in the developing world. Since much of this direct investment offered by China and Russia is made through direct aid or heavily state-subsidized corporate (i.e., private sector) interests, the United States is losing influence in these nations.

Accordingly, in the era of Great Power Competition, considering private sector interests as a force multiplier to meet national strategic aims is worthy of further discussion; particularly if interests can be aligned in a way that improves the quality of life for vulnerable populations in the affected state and precludes the severity of future outbreaks.

Insight four: differences in assuming risk

Risk is perceived differently by different stakeholders. For example, the NGOs and IGOs represented in this wargame, many of whom were frontline responders, possessed a very different perspective of risk than the military players. According to the participants representing NGOs and IGOs in Urban Outbreak 2019, they were willing to assume a high degree of personal risk if required to save lives and alleviate human suffering. In contrast, military forces assessed risk in terms of potential harm to their own operational readiness to accomplish specified missions such as security, logistics, and sustainment. The risk lines were blurrier for military medical personnel, who tended to lean more into assuming personal risk (more akin to the NGOs/IGOs) than their non-medical military peers.

Insight five: stovepiping and cultural resistance to change and adaptation

The Urban Outbreak 2019 Wargame brought together a wide range of organizations with specialized skills and talents engaged in humanitarian response. However, these entities were not necessarily able to work across constituencies. The largest bureaucratic organizations presented the most inability to change or adapt to conditions unfolding on the ground. The more complex the environment, the less flexible larger systems appeared to be in meeting these challenges. Specialized and siloed expertize invited failure, particularly during the second move of the game where adaptable approaches to working across multiple sectors were necessary to mitigate further propagation of the outbreak within Olympia and surrounding environs.

Lessons learned from Urban Outbreak 2019 and its applicability to America's COVID-19 response

While wargaming is not experimentation, and its findings are not predictive, the applicability of the insights derived through Urban Outbreak 2019 game play are informative to senior decision-makers engaged in COVID-19 response. As previously discussed, Urban Outbreak 2019 as a gaming construct was conceived long before COVID-19 presented itself on the world stage, and it was approximately six months before the full weight of this pandemic was felt outside Asia in the United States, Europe, and South America. Accordingly, it may seem odd that the findings of a notional game focused on a readily treatable pathogen impacting a fictitious mega-city in a developing nation would yield valuable lessons for America's leaders responding to COVID-19. However, the five insights discussed in the previous section of this chapter each have meaning, because there is no one, cohesive national strategy for addressing COVID-19; but rather a complex serious of micro-level relationships and intra-relationships between municipalities and states.

This lack of unified, centralized national response within the United States means that many states' limited public health systems are more akin to an Olympia Ministry of Health and its modest local emergency medical response capabilities than a more robust capacity proffered by the Centers for Disease Control, National Center for Disaster Medicine and Public Health, or Federal Emergency Management Agency. Given the limitations of this state and local patchwork approach to addressing COVID-19, rather than through a centralized framework suggests that each of the five key insights identified in the Urban Outbreak 2019 wargame is magnified substantially due to this suboptimized solution to mitigating the spread of this virus.

The admonition manifest from Urban Outbreak 2019 game play that even a slight perturbation within one system of a sector may have impacts across other sectors is particularly troubling given the fragility of the inter- and intra-state logistics in the current US-based response to COVID-19. For example, an unanticipated spike in cases in a municipality or state may create enough churn to disrupt regional supply chains impacting everything from food to medicine. Many states have already seen dramatic rises in food insecurity among their populations due to peaks earlier in the spring of 2020, and with a second wave of infection anticipated in the fall, similar shortages could prove catastrophic for vulnerable populations such as children, the elderly, and immunocompromised.

Although some efforts have been made to bring the private sector into both the initial and crisis phases of COVID-19 response far too little emphasis has been placed on leveraging the Defense Production Act of 1950 to support the development of medical and public health infrastructure and equipment. Given that a national response framework has been eschewed in deference to a jumble of individual state and micro-regional compacts, it is unlikely that the President of the United States will call for further unity of effort on the part of America's industrial base—save perhaps for the pharmaceutical industry in its hasty quest to bring a viable vaccine to market through Operation Warp Speed.

It is also unlikely that greater use of federal emergency response assets and personnel for the purposes of contact tracing will be deployed. This is unfortunate as well. For example, the robust capacity for medical surveillance and health monitoring performed by the Occupational Safety and Health Administration and the National Institute for Occupational Safety and Health could be a game-changer. By deploying industrial hygienists and health and safety investigators to develop a common operating picture and promulgate commonsense personal protective equipment standards for known industries, America would be able to mitigate hazards in specific sectors until such time as a vaccine is readily available.

Conclusions

The less structured and indeterminate the problem, the more powerful and useful wargaming proves in identifying unknown issues and gaps in thinking and exposing cognitive biases. Moreover, the use of this wargaming approach can reap great benefits for senior decision-makers interested in uncovering the second- and third-order effects of their actions.

No environment is better suited to apply wargaming's power than the urban landscape of a megacity, where complex issues involving disparate cultures, socioeconomic inequality, and inequities in access to health care, sanitation, protection, education, and a host of other catalysts to instability abound. Through the gaming process, leaders from a variety of agencies, organizations, and interests can participate in a shared event, which allows them to identify gaps, overlaps, and seams.

Paraphrasing Lieutenant William McCarty Little, by engaging urban stakeholders in robust and lively conversation, grounded in a common experience, "shifts in beliefs, judgments, or actions occur that prove the most valuable in garnering insights useful to senior decision makers" [1].

References

[1] Brightman HJ, Dewey MK. Trends in modern war gaming: the art of conversation. Nav War Coll Rev 2014;16:17–30.

[2] Burns SW. War gamer's handbook: a guide for professional gamers. Newport: Defense Automation and Production Service; 2013.

[3] Davies B, Brightman H, Brosteun J, Polatty D, Card B. Urban Outbreak 2019 pre-analytic "quick look". Newport. Defense Automation and Production Service; 2019.

[4] Davies B, Lovett KR, Card B, Polatty D. Urban outbreak 2019 pandemic response: select research & game findings. Newport: Defense Automation and Production Service; 2020.

[5] Brightman HJ. Urban outbreak 2019 wargame data collection & analysis plan. Newport: Naval War College. Civilian-Military Humanitarian Response Program; 2019.

Chapter 10

The adaptability and resilience of cities to major epidemics: a comparison of Sydney and Phoenix subject to a hypothetical smallpox epidemic

David James Heslop[1], Raina Chandini MacIntyre[2], Brian Gerber[3]
[1]School of Public Health, University of New South Wales, Sydney, NSW, Australia; [2]The Kirby Institute, University of New South Wales, Sydney, NSW, Australia; [3]Watts College of Public Service & Community Solutions, Arizona State University, Phoenix, United States

Since antiquity, human populations have moved from a predominantly agrarian subsistence existence toward larger and larger groupings of individuals (villages and towns) resulting ultimately in the emergence of the city [1]. As cities emerged subgroups of city populations segregated into specialized cohorts of individuals for necessary or desirable cultural, economic, or political reasons [2,3]. While on the one hand this generated advantages and benefits for the city, enabling its survival, growth, and prosperity, it also created economic and resource dependencies between individuals and groups, and also significant vulnerabilities [3]. One such vulnerability was the propensity of cities to facilitate the occurrence of outbreaks of infectious disease [4,5].

Innovation, cultural change, and social change were necessary imposts placed on emerging city populations and their broader societies, further fueling adaptation and new dependencies, and requiring the behavior, activities, and daily pattern of life of city populations to evolve. In modern cities, these same interrelationships between city form, function, population, culture and behavior, and economic activity still exist. For the resident, they become a foundational and often imperceptible fabric of city life, and are the "hidden hand" influencing and driving the short- and long-term behaviors and choices

Inoculating Cities. https://doi.org/10.1016/B978-0-12-820204-3.00010-3

of individuals. More recent technological advances since the industrial revolution have fueled the emergence of an accelerating number of interdependencies, behavioral and cultural changes in city populations, but have equally enabled rapid population growth in many cities driven by advances such as sanitation, hygiene, modern medicine, and public health. As these cities grow, the basic structure and operation of the city have scaled up and developed complex social and technical interrelationships among the larger population groupings.

In this chapter the relationship between a city's population, its form, structure and function, and the environment in which the population lives and interacts will be explored in the context of a major epidemic threat − smallpox, a pathogen that has historically accounted for large epidemics with high mortality. Sydney, Australia, and Phoenix, United States are used as case studies to illustrate how social, economic, health, and other factors are important determinants of predicted outcomes in these cities were they subject to a smallpox outbreak.

Cities, populations, and pathogens

Cities are complex human-centric environments that have evolved significantly over human social development and history. The social, economic, political, design, physical environment, geographic, meteorological, and many other dimensions that make up the actuality of the physical and perceived city environment interact strongly with the social, cultural, economic, and political activities of the population. In other words, the form and operation of the city as an entity is influenced by and interacts with the behavior, decisions, and activities conducted by the individuals making up the city population.

Since the 1920s the dynamics of how diseases spread within a population have been understood mathematically − frequently with excellent approximation to observed epidemic data [6]. However, these models often assume unrealistic simplicity in the demographics, compartmentalization, subgroup interactions, and personal contacts of individuals within those populations. While at a population level, the effects of compartments and subpopulations are smoothed out; in population contexts, where high-intensity contacts or alternative routes of exposure are possible, these may not be properly accounted for. The complexity of cities allows for combinations and dynamics that simple models cannot properly account for (Fig. 10.1). Individuals interact with various components, objects, materials, and surfaces of the built environment of a generic city in highly disparate ways over time, alongside interacting with diverse subpopulations of individuals. Each individual's interaction with the environment and population determines how transmission of specific pathogens may or may not occur in the city. The complexity, frequency, and nature of these interactions are distinctly different from other human social environments (e.g., rural environments).

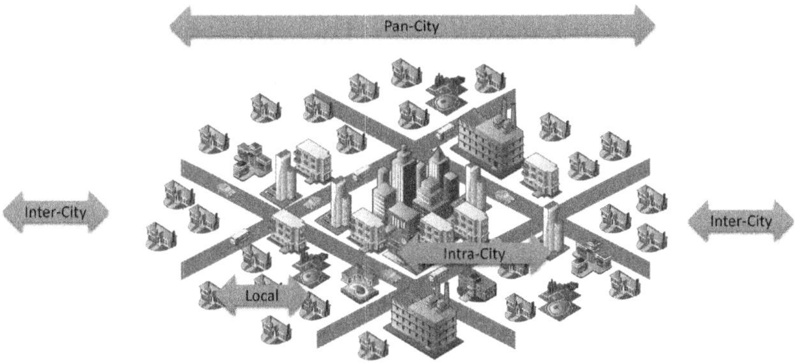

FIG. 10.1 Individual population interactions within a typical city. Individuals will interact with various components, surfaces, and objects making up the built environment, and different population subgroups. This is replicated across many different layers of interaction types, social strata resulting in complex and at times chaotic interactions over time.

Infectious disease interaction with urban settings

The genesis of epidemics in populations in cities

At their most basic, epidemics emerge as a consequence of the interaction between individuals and pathogens, mediated through an environment that modifies the nature of transmission events. This has historically been understood through the perspective of the "host-environment-pathogen" model familiar to public and population health specialists [7]. However, hazard exposure models — such as the "source-pathway-receptor" model familiar to occupational and environmental medicine and occupational hygiene specialists — provide additional and complementary insights that are relevant to disease transmission [8–11]. Understanding how pathogens dispose or contaminate an environment while also understanding how a pathogen spreads between individuals in populations living in that environment must draw from both theoretical frameworks.

Exposure dynamics relevant to disease spread

Individuals in a population can be exposed to a biological hazard, in this case, a pathogen, through four fundamental routes of exposure (Table 10.1) [12].

Beyond these broad routes of exposure, the way in which the pathogen and its carrier medium (e.g. aerosols or respiratory droplets) interact with the functional anatomy and physiology of the human body allows more detailed classifications of hazard transmission capability, such as airborne, droplet, fomite, or secondary aerosolization (Table 10.2).

The many infections already within cities

Populations within cities are constantly subjected to infections of many different types, both biological and non-biological, endemic, and epidemic.

TABLE 10.1 Fundamental human routes of exposure to biological hazards.

Route of exposure	Definition
Respiratory	The inhalation of, or introduction of, pathogen into the airways or lungs
Ingestion	The introduction of pathogen into the digestive tract
Dermal	The introduction of pathogen into the body across the dermis
Mucosal	The introduction of pathogen into the body through mucosa

TABLE 10.2 Pathogen transmission characteristics and capabilities.

Route of exposure	Hazard capability	Definition
Respiratory	Airborne transmission	Transmission by respiratory route by extremely small aerosolized droplets, or in the absence of droplets.
Respiratory	Droplet transmission	Transmission by the respiratory route only by aerosolized droplets > 5 μm in diameter.
Any	Fomite transmission	Transmission by any route of exposure whereby a contaminated intermediary object allows transfer between individuals
Any	Secondary aerosolization	Pathogen that has settled on a surface is reintroduced into the air via various kinetic means (wind, passage of individuals) with the potential for transmission to other individuals.

Fundamentally an epidemic is the sustained transmission and spread of an item within a population — skills, knowledge, rumors, preferences, and pathogens. For an infectious disease outbreak to be classed as an epidemic, a population must see a rapid increase in incidence above a baseline threshold within a geographically defined area. This is distinct from endemic diseases where case incidence does not deviate from the baseline. Mathematically an epidemic is often defined where the factor R0 is greater than 1 [6], which gives rise to epidemic conditions and the potential for exponential growth. This occurs as a consequence of the unique characteristics and relationships that enable the transmission to occur and related to the host population, the environment, and the features of the pathogen itself. At any one time in most cities around the world, there are multiple concurrent outbreaks of common respiratory pathogens — such as enterovirus, influenza, rhinovirus, and many others. There

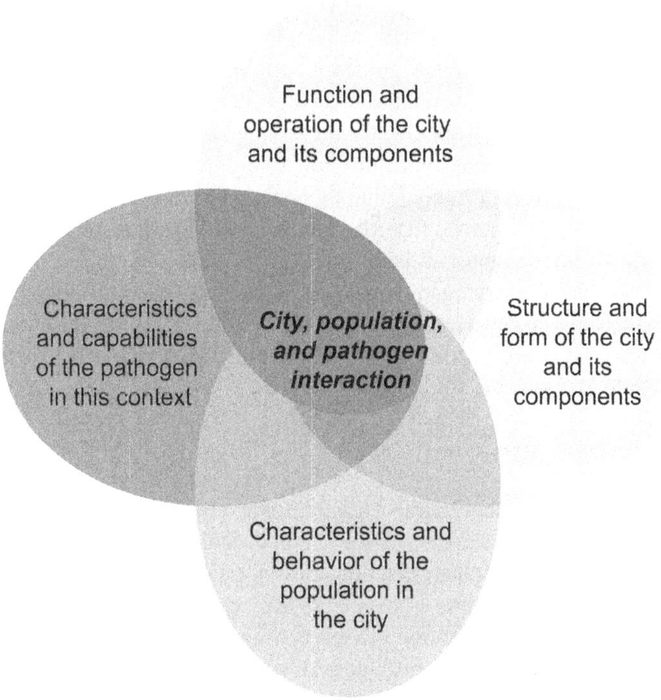

Function and
operation of the city
and its components

Characteristics
and capabilities
of the pathogen
in this context

*City, population,
and pathogen
interaction*

Structure and
form of the city
and its
components

Characteristics and
behavior of the
population in
the city

FIG. 10.2 Primary factors influencing the city, population, and pathogen interaction.

are also endemic outbreaks, not necessarily caused by respiratory pathogens, such as malaria and HIV.

Cities and their populations are adapted and habituated to minor epidemics, which together result in a disease burden that is managed within the existing health system capabilities and capacities of the city. But the emergence of a new pathogen within a city's population can disrupt this delicate balance and result in an epidemic. This, in turn, necessitates the operation of the city and behavior of the population to adapt and change. The capacity, capability, and willingness of the population, constrained by the physical and operational realities and flexibilities inherent to the operation and structure of the city, are the primary determinants of the ability of the city and its population to manage the unfolding crisis (Fig. 10.2).

Operational flexibility and adaptability of cities

Flexibility and adaptability to an unfolding epidemic are important determinants in the ability of that city and its population can do to minimize the consequences of an event. While a city may have the ability to close down various aspects of its social, economic, and public services and structures for a time, this will ultimately impact the economic activity and other essential city functions. Equally some cities will be unable to adapt to changing circumstances in the timeframe of the epidemic, limiting the options available to decision-makers and the population. Under certain conditions, the very structure, operation, and form of the city may increase the intensity and impact, making the outbreak more difficult to control. Stated another way, a principles-based approach to epidemic management is important as each city will have unique requirements, features, and contexts that demand case by case management of the epidemic at hand.

Below, we broadly aggregate some of determinants that are important for understanding how an epidemic might unfold in a city, taking into consideration key social, economic, demographic and behavioral patterns of the population (Table 10.3).

Smallpox

Smallpox is caused by the variola virus and was eradicated in 1980. Known seed stock of the virus exists in the United States and Russia, and the destruction of these stocks has been debated for decades. However, the risk of re-emergence of smallpox has arisen because of synthetic biology [13], and it ranks as highly as anthrax among category A agents. Indeed, in 2017, Canadian scientists synthesized an orthopoxvirus, effectively demonstrating that synthesis of the virus in a laboratory is possible [14].

Smallpox is a respiratory disease and primarily spread through the airborne respiratory route, although spread by contact and fomites has also been documented [15]. Very long-range transmission, termed aerial convection, has also been described. The phenomenon was first recognized in London, where it was observed that secondary cases occurred in the community within 1−2 km proximity to smallpox hospitals. Therefore, the density of living in urban environments may be influential in the spread of smallpox [15].

The disease causes a rash and fever, and in a small percentage may be hemorrhagic. The R0 is probably between 4 and 5, meaning that, on average, every one case of smallpox would result in between 4 and 5 subsequent cases [16]. In addition, immunosuppression due to advances in medicine and people living with HIV has resulted in nearly one in five people in cities such as Sydney, living with moderate to severe immunosuppression [16]. This may result in a higher case-fatality rate than the 30% described in the pre-eradication era.

The fundamental components of an effective response to a smallpox attack scenario would include case finding, contact tracing, isolation, quarantine, ring or mass vaccination strategies, social distancing, travel restrictions, and face masks. In an attack scenario, any delay in the response or failure to achieve high rates of case finding, contact tracing, isolation, and quarantine could result in a large smallpox epidemic [17]. Of particular note, in a study of health

TABLE 10.3 Determinants of epidemic spread.

Domain	Determinant
Pathogen	Infectivity Mode(s) of transmission Environmental persistence Severity of illness Seroconversion rate and effectiveness
Interpersonal contacts	Contact frequency between individuals Social distancing Gathering events (i.e., mass and minor) Outdoor activities Recreation and leisure activities
Population flexibility and adaptability	Individual self-quarantine capacity Nature and type of essential activities Nature and type of non-essential activities Household demographic and social structure Population ability to self-isolate Flexibility regarding important events (i.e., religious, political, demonstrations) Funerals
Transport	Type and utilization of mass transit Individual transport options
Special groups	Homeless Disabled and infirm Mentally ill Admitted patients Palliative care Epidemic isolated and quarantined Unaccompanied minors
Economic	Population income security Population savings/financial resources Employment security Child care options Aged care options School opening and closures
Health	Health facility numbers and types Health workforce numbers and types Health system redundancy, flexibility, and surge capacity Pharmaceutical supply and sustainment Role and function of primary and secondary care
Health — epidemic specific	Specialized pandemic management facilities (e.g. fever clinics, recovery centers, etc.) Safe and dignified palliative care Intensive care surge capacity Health logistic capacity and capability

Continued

TABLE 10.3 Determinants of epidemic spread.—cont'd

Domain	Determinant
	Sustainment of essential health functions (essential surgery, obstetrics, others) Personal protective equipment
Built environment	Residence types (e.g., detached houses, apartments, etc.) Workplace design and dynamics Pattern of population densities in built environment zones Physico-chemical characteristics of the built environment Interactions between environment and population behavior (e.g., high-touch objects)
Workplace	Workplace dynamics Work ethos and expectations of work Work location flexibility Employment type (primary, secondary, tertiary)
Natural environment	Climatic conditions Geographic features
Basic needs and essential services	Water supply Electricity supply Heating supplies Food and food preparation Emergency shelter Essential personal hygiene supplies Emergency security response Emergency health response Emergency legal needs Emergency administrative needs Mortuary affairs
Waste and sanitation	Sanitation system Waste system Decontamination services
Government	Agile command and control structures Inter-agency coordination Emergency powers and legislation Review and audit

system capacity in Sydney, contact tracing capacity and hospital beds could be exceeded rapidly if the response is delayed or incomplete [18].

Vaccination for smallpox is highly effective and was integral to eradication efforts. Prior to eradication, the cities discussed in this chapter, Sydney and Phoenix, had different smallpox vaccination policies. In Australia, mass vaccination was never used, with only about 25% of the population, mostly migrants, having past vaccination. In the United States, in contrast, mass vaccination was used, so that most people aged over 40 years would have been vaccinated [16]. Still, the influence of past vaccination is minimal, as vaccine-induced immunity wanes substantially after 10 years [19].

Further, if an engineered virus is created that is vaccine-resistant, this may not be the case. The US National stockpile has enough smallpox vaccine for the whole population. In Australia, however, the stockpile is limited. Antivirals against smallpox are also available, with TPOXX being licensed by the Food and Drug Administration in the United States. These are important for treatment, but may are not indicated for prevention. The control of smallpox in a city would require human resources, physical resources (including spaces for quarantine and isolation), vaccine, drugs and coordination of the response, as we showed in a simulated smallpox pandemic [20]. We illustrate below how this may vary depending on the context, using Sydney and Phoenix as case studies.

Case study: Sydney, Australia

Description and features

Sydney is the largest city in Australia and had an estimated resident population of around 5,312,200 persons in June 2019 [21]. While the country has a very low overall population density at 3.3 per square kilometer, Australia is one of the most urbanized countries, globally. Compared to other cities in Australia, Sydney has the largest combined area of high and very high-density areas (i.e., 187 square kilometers where population density is more than 5000 persons per square kilometer). However, these high-density areas accounted for only 1.5% of the total area of Greater Sydney. In contrast, 31.5% of the area of Greater Sydney has a population density of fewer than 500 persons per square kilometer (Fig. 10.3) [22].

In 2016, a majority of individuals in Sydney resided in low-density accommodation [23]. This is characterized by single or double story detached residences, often on a larger block of land. Recently there has been a significant increase in apartment construction, but these are clustered in specifically planned areas — often close to employment, industrial, transport links, and other resources. In inner parts of the city, older apartment stocks are characterized by small detached blocks of apartments of no more than 4 levels.

Demographics

Sydney has an aging population (Fig. 10.4). The largest age group in both genders is the 30—34 year cohort. 13.85% of the population is aged 65 and above, while only 12% of the population is aged 0—19 years of age. There is marked variation in average age by region within the city, with high growth areas of the South West, West, and North West being characterized by a much

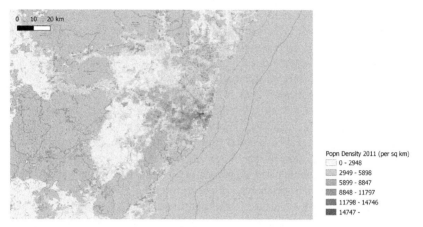

FIG. 10.3 Population density of Greater Sydney Area [22].

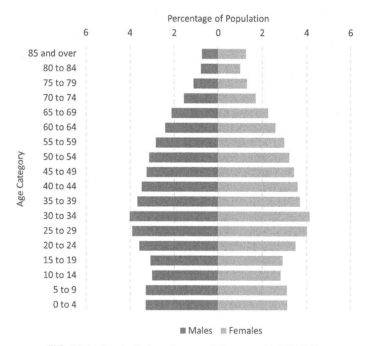

FIG. 10.4 Greater Sydney Area population pyramid, 2016 [23].

lower average age, whereas more established areas closer to the city of Sydney have a greater proportion of older residents.

The average household size in Australia is 2.6 individuals [23]. Households are generally comprised of single or partnered adults, or two-generation households (i.e., parents and children). In Australia, just under 17% of households were classified as multi-generational (i.e., multiple adult generations in one household) in 2012, with the remainder comprising either single households, partners, or parent-children households [24]. The size of the average detached residence in Australia is also large by international standards at 230 square meters of floor area in 2018 [25]. It is also unusual for children to co-habit the same sleeping area with parents from school age, and uncommon for older children to co-habit the same sleeping area with siblings after entering high school from around the age of 12.

Socioeconomic conditions

There is substantial socio-economic variation across the Greater Sydney region. Areas close to economic centers of activity such as the Central Business District emerging regional economic hubs within the Sydney region have higher levels of employment, average income, educational options and attainment, and access to a wide variety of resources. The median yearly household income in Sydney in 2016 was 91,000 AUD (67,000 USD) [23]. Areas distant from the Central Business District are characterized by lower levels of employment, inter-generational unemployment, higher reliance on social welfare, and more limited access to education, training, and other resources. Broadly, the North and East of Sydney is considered to be affluent, whereas the West and South West are considered to be less affluent.

Health system

There are ten major (i.e., tertiary referral) hospitals within the Greater Sydney region, and a total of 12 major hospitals across the State of New South Wales [26]. Each of these facilities is the primary referral facility for their respective Local Health Districts, from 17 feeder hospitals that provide fewer acute or specialized services. Two additional specialist pediatric hospitals are also located in the Greater Sydney Area.

The nine major hospitals are not equally distributed uniformly across the Sydney region, with many being located in early economic development areas close to the city of Sydney. The West and South-West of Sydney are relatively underserved compared to the East and North areas of the city. This is also reflected in variation in the accessibility to primary care, allied health, private healthcare, and other health services, and correlates closely with variation in socio-economic status across the Sydney region. Nevertheless, recently published data suggest that despite the geographic variation in health care accessibility, key indicators such as travel time to hospitals are comparable for most areas of the city [27].

Essential and emergency services

Essential emergency services in Sydney (e.g., police, fire and rescue, ambulatory, and emergency welfare services) are predominantly under the control of and funded by the State Government. As such there are significant constraints on surge capacity and flexibility, with all services tailored to meet business as usual with limited additional capacity on demand. Essential services such as water supply, power, and food supply are either partially under government control or privatized. In Sydney, these services are reliable and high quality, with significant surge capacity under most conditions including crises such as drought, heatwave, or fire.

Preparedness

Preparedness for emergencies in Sydney is influenced by federal (i.e., national), state, and local activities. In Australia, the responsibility for planning and preparing for, responding to, and recovering from health emergencies falls predominantly on the state and local governments. Organizations at the federal level — most notably the government departments of Health, Home Affairs, and Defence and their specialist sub-organizations, such as Emergency Management Australia — provide oversight, additional resources on request, and overarching command and coordination mechanisms. A key feature of the Australian system is the quasi-independence of the states whereby federal assistance can only be delivered on request under normal circumstances, and unilateral intervention can only occur in extremis.

In New South Wales, the framework for planning, preparedness, response, and recovery is outlined in the State Emergency Management Plan [28]. This document outlines the key responsibilities for different types of disasters and events. The lead agency is known as the "combat agency" and reflects the key front-line role of that organization. For pandemics, the Health Department is the designated combat agency; however, in almost all circumstances the New South Wales Police Force provides centralized command, control, and communications capabilities. Subordinate to the State Emergency Management Plan are the Biosecurity (Animal and Plan) Emergency Management and Human Pandemic Influenza Plans [29,30]. These plans provide detailed frameworks for how prepared resources will be marshaled and utilized in the event of a major biosecurity event. Complementing these plans are logistic support organizations, such as HealthShare [31], that provide centralized coordination of essential support services during pandemic events.

Vulnerabilities

The high degree of dispersal of the Sydney population results in a number of community effects. At the individual level, many individuals in Sydney are dislocated from each other. Community interaction is limited, with up to half of the residents reporting feeling disconnected even in the most densely populated parts of the city. Equally, loneliness and the consequences related to loss of long-term partnership over time have been reported as a significant problem, particularly in the elderly [24].

Sydney, and Australia more generally, is located far from key global manufacturing areas. This reality creates vulnerabilities related to the supply of critical preparedness and response items. For large epidemics, the availability of personal protective equipment (PPE), essential medicines, testing supplies and equipment, key medical equipment (e.g., ventilators) are only accessible at the end of extended lines of supply. These are vulnerable to supply delays, price elevation and gauging, and impacts of limited purchasing size and power due to the limited Australian market size. These effects have been acutely felt in the recent coronavirus pandemic. While local manufacturing could potentially meet demand, the commercial viability and sustainability of local manufacturing are equally problematic. Stockpiling is also a consideration. In December 2019, Australia experience among the worst bushfire season documented [32]. As a result, an already limited national medical PPE stockpile was depleted — leaving the country short of masks and respirators when the COVID-19 pandemic began. In 2019, modeling showed that an epidemic of smallpox ongoing for six months would require at least 30 million respirators for Sydney health workers alone, but when the bushfires hit, less than 10% of that existed in the stockpile [33]. The impact of multiple serial disasters is rarely accounted for in planning.

The health system in Sydney is also adapted to manage the existing health burdens confronting the city's population. This is focused on the management of chronic diseases, predominantly in older populations. Due to financial and resource pressures, the system is finely tuned to minimize waste and often operates at its full operating capacity. Were there to be a major outbreak of smallpox in Sydney — which would represent a significant deviation from the existing health burden — a rapid adaptation and surge requirement would be generated. While some health burdens would potentially decrease (e.g., road trauma) this would likely be offset by a large increase in outbreak-related clinical presentations. The repurposing of existing facilities, the formation of "fever hospitals" for infectious disease patients only, and the establishment of expanded intensive care unit capacity would likely be required. This would almost certainly impact the routine activities of the healthcare system and probably require non-essential medical care to be canceled or deferred. The impacts of such actions — particularly the consequences of delayed chronic disease or semi-urgent surgical care — would emerge and add to the burden of disease associated with any major outbreak.

Summary: Greater Sydney Area

While Sydney is a large city of over 5 million people, and it has significant resources at its disposal for the management of an epidemic, an outbreak of smallpox would significantly challenge the entire population and its support structures. With sufficient warning and time, the adaptation of aspects of the physical built environment alongside focused behavioral, cultural, social, and economic changes in the population, would allow Sydney to adapt to an unfolding crisis. Vulnerabilities specific to Sydney and its population include limitations on economic adaptability and resilience due to high household debt, limited surge capacity in the health system beyond business as usual activities, and the geographic dispersal of the population. Resilience factors specific to Sydney include generally low-density housing stock with the ability for households to isolate, high degrees of financial and personal resources for many in the population, and a strong social welfare system focused on vulnerable subgroups.

Case study: Phoenix, Arizona, United States of America

Description and features

The city of Phoenix is the largest in the state of Arizona with about 1.7 million residents. It is located within Maricopa County, which itself has about 4.5 million residents, representing the fourth largest county in the United States. Maricopa County has seen steady and significant population growth; in raw numeric terms, it grew by about 700,000 residents from 2010 to 2019 [34]. The basic spatial distribution of Maricopa County's population is similar to Sydney, Australia (Fig. 10.5). The city of Phoenix covers most of the highest population density area within the county on the basis of persons per square mile (relative to the two dozen cities within the county jurisdiction). As the figure shows, the county's population density declines steadily with movement away from the city to the outer areas of the county's jurisdictional boundaries—a spatial distribution pattern similar to the Greater Sydney Area.

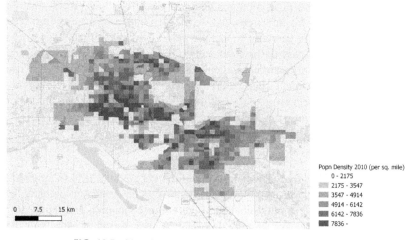

Popn Density 2010 (per sq. mile)
0 - 2175
2175 - 3547
3547 - 4914
4914 - 6142
6142 - 7836
7836 -

0   7.5   15 km

FIG. 10.5 Phoenix metro area density, Maricopa County, AZ.

Roughly 66% of Maricopa County residents reside in single-family detached housing structures residences, or low-density housing [35]. Roughly ten percent live in medium-density housing (i.e., between two and nine units per housing structure) and about fifteen percent can be identified as residing in high-density housing structures (i.e., 10 or more units per housing structure). The Arizona Department of Housing reports that in 2006, housing construction was distributed as about 85% single-family and 15% multi-family structures. But there is a noticeable trend toward higher density living arrangement: just over a decade later, in 2017, the share of multi-family (medium- or high-density housing) had about doubled to represent about 29% of new construction as being medium or high density, and the remaining 71% were single-family housing units [35].

Demographics

The population of the greater Phoenix metropolitan area (indicated roughly as Maricopa County overall) is slightly older than Sydney: 14.4% of the population is aged 65 and above (Fig. 10.6). However, it also has a much higher proportion than Sydney of residents who are aged 19 years or younger — 27.2% of the county's population. Overall the median age in Maricopa County is 36.1 years old, and the county's largest age group is 20–25 (14.5% of the

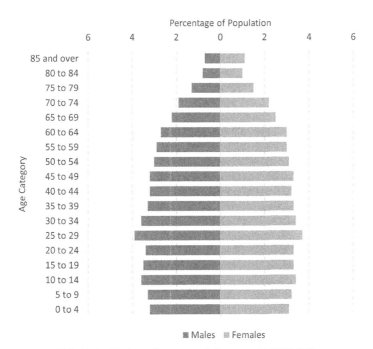

FIG. 10.6 Maricopa County population pyramid, 2019 [36].

county's residents). This age demographic pattern indicates some degree of regular inflow of younger new residents coinciding with a stable elderly population. Phoenix's population pyramid shows a pattern dissimilar to Sydney, which tends to have smaller numbers at the high and low range of the age scale than does the Greater Phoenix Area [36].

When considering the spatial distribution of age cohorts in Maricopa County overall, there are also dissimilarities when comparing to Sydney. A majority of the over 60 age cohort live on the outer edges of the Phoenix metro area while a majority of younger cohorts (i.e., young- and middle-aged persons) reside within Phoenix city limits and closer to central population areas [37].

The average household size in Maricopa County is 2.8 persons; about 48% of households include married couples. A large majority of households in Maricopa County, 70%, do not have children. Household size and generational arrangements in living arrangements in Maricopa County are similar to that of Greater Sydney: large family and multi-generational households are uncommon.

In terms of racial or ethnic diversity, Phoenix and Maricopa County are more diverse than the United States as a whole. About 56% of Maricopa County identifies as non-Hispanic White and about 31% identifies as Hispanic. African Americans make up just over five percent of the county population, Asia-Americans comprise just over four percent and Native Americans represent just under two percent of the population [38]. One implication of this diverse population is a greater accompanying linguistic diversity than is true in comparison to the US overall. About 27% of Maricopa County residents speak a language other than English — and a plurality of those individuals indicate some degree of limited English language proficiency.

In terms of the distribution of educational attainment in Maricopa County, a significant proportion of adult residents (i.e., measured as 25 years of age or older), nearly 13%, have not completed high school and earned a degree. Another 23% completed high school while another 24% have just some amount of undergraduate education — but without completing a degree. Of the remainder of the population, about nine percent have an Associate's Degree and the other 20% completed a Bachelor's degree program. Finally, 12% of county residents have earned a graduate or professional degree [39].

Socioeconomic conditions

The median household income in Maricopa County is 61,000 USD (89,000 AUD), which is slightly below the national median in the United States. About 15% of Maricopa County households fall below the poverty level. The relevant spatial distribution for this indicator of socioeconomic status is at least somewhat associated with age; the distribution of older residents tend to be higher median income areas whole the central areas of the county, the core areas of the city of Phoenix tend to have lower median incomes. At the same time, the spatial distribution of median income tends to be lower in the exurban, lower population, rural or desert municipalities and townships in Maricopa County, especially in the southern part of the county. The northern suburbs of Phoenix tend to be more affluent compared to other parts of the county [40].

Health system

In strictly local terms (i.e., excluding discussion of broader national health policy or healthcare delivery systems) considerations pertaining to the delivery of health services and resources might be most conveniently summarized with reference to the hospital sector in Maricopa County. The Arizona Department of Health Services indicates there are 75 hospitals in the county, with about one-third of those being large, high-resource capacity hospitals. Of the larger facilities, nearly all are located in the city of Phoenix or proximate to the city's boundaries. The county has a collection of specialty private hospitals as well, some of which are located in the less densely, rural southern area of the jurisdiction. Of those hospitals listed as licensed by the Arizona Department of Health Services, only one is classified as a public hospital. The other licensed hospitals are either classified as having private corporate or non-profit organizations ownership structures [41]. Maricopa County has two pediatric facilities, both part of the Phoenix Children's Health System. The remaining facilities in the county are generally smaller branch facilities, designed to integrate as health resource centers serving specific community areas.

Essential and emergency services

A common aphorism for describing the management of emergencies and disasters in the United States is that local governments execute operations, state governments manage them, and the federal government supports them. While this is a grand simplification, the expression does convey appropriately that local governments (i.e., city, county, and other sub-state jurisdictional entities) are essential operational actors who take the lead in providing personnel responsible for work in the field during an emergency or disaster. Like any other county government in the United States, Maricopa is a county governmental body itself and as such, key emergency management and public health units (e.g. the Maricopa County Sheriff's Office, the Maricopa Department of Emergency Management, the Maricopa Department of Public Health) constitute some of its administrative agencies. However, an important response services note is that it does not provide directly county-wide public emergency medical or fire services. While county emergency management and public health agencies perform key functions, municipalities typically fund and provide law enforcement, fire, and emergency medical services directly. Or such services are provided through partnerships or coverage agreements with public agencies or private vendors for fire or ambulance services. Likewise, those same municipalities within the county establish formal mutual aid agreements to provide horizontal coordination for maintaining emergency response and public health services and coverage throughout the various jurisdictions in the county, a practice that is commonplace among local governments in the United States [42,43].

Beyond public health, medical, emergency or first response resources, other essential services such as water and electricity resources are either provided by utility cooperative and government-owned corporations privately held utility providers. In general terms, the utilities sector is regulated by the State of Arizona Corporation Commission.

Preparedness

Preparedness for emergencies and disasters in Phoenix and Maricopa County, like other local governments in the US, is critically shaped by the over-arching federal governance structure. This is true for the broad range of response and incident command protocols developed after the September 11, 2001 terror attacks and the broader National Response Framework (NRF) [44,45]. Within the NRF, various bio-threats are considered and addressed through guidance provided under Emergency Support Function 8, Health and Medical Services, which details how the U.S. Department of Health and Human Services, Office of Preparedness and Response, helps lead the management and coordination of the federal medical response to major emergencies and disasters — including those arising from natural and tech-nological hazards, as well as acts of terrorism, which would be the most likely of a smallpox event in Phoenix.

Under that rubric of federal resources for medical response to disaster is the National Disaster Medical System (NDMS) [46]. The NDMS is defined as:

A federally coordinated system that augments the Nation's medical response capability. The overall purpose of the NDMS is to supplement an integrated National medical response capability for assisting State and local authorities in dealing with the medical impacts of major peacetime disasters and to provide support to the military and the Department of Veterans Affairs medical systems in caring for casualties evacuated back to the U.S. from overseas armed con-ventional conflicts.

These federal-level bureaucratic resources and strategy and operational guidance documents are also shaped by key statutes with specific relevance to epidemic and pandemic incidents, such as the 2006 Pandemic and All-Hazards Preparedness Act (Public Law No. 109-417), which among other things lays out a framework for the development of, and system for, acquisition and deployment of medical countermeasures. While that statute was reauthorized in 2013, the U.S. Department of Health and Human Services released an updated pandemic influenza planning document to shape national response efforts and in 2018, the White House Office of the President released a National Biodefense Strategy for the purposes of "strengthening the biodefense enterprise, [and] establishing a layered risk management approach to countering biological threats and incidents" [47].

In other words, a set of federal statutes and federal documents enunciating strategic doctrine over hazards and disaster management in general sets a framework for more specific risk management approaches relevant to epidemics and pandemics. Tools and guidance produced at the national-level are then essentially replicated at the state- and local-level. In Arizona, the state uses a response and recovery plan parallel to the NRF, including a set of delineated responsibilities associated with a set of emergency support functions [48]. In turn, specific local actors, including the Phoenix Office of Homeland Security and Emergency Management, the Maricopa County Public Health and Emergency Management offices, are the primary agencies for leading a coordinated epidemic response.

Vulnerabilities

From the perspective of Phoenix and Maricopa County, vulnerabilities specific to epidemics fall into several broad categories, including the quality of operational relationships to state and federal actors, planning adequacy, resource capacity, and community characteristics and likely behavior.

Given public health and emergency management agencies at the city and county level work closely with state and federal government, and such relationships are central to guiding response operations, assessing the robustness of those relationships is critical to any vulnerabilities assessment. In an early review of the national system response to the COVID-19 pandemic, federal evaluators found a range of significant intergovernmental coordination challenges across the multiple dimensions of incident management, beyond narrow public health concerns to include short and long-term economic impacts as well [49]. As the report indicates, the public health challenges at an epidemic or pandemic scale necessitate effective national and sub-national coordination to reduce risk and vulnerability.

Moving from a national perspective to local resource capacity illustrates this point. Phoenix and Maricopa have robust and well-staffed emergency management and public health agencies. However, the operative question is how well-resourced those agencies are in the face of a short-term versus a prolonged epidemic. From a short-term response and overall planning perspective, a well-defined state emergency response plan [48], coupled with local resources suggests sufficient resources to cope effectively. However, as the COVID-19 pandemic response unfolds, early assessments indicate that poor political leadership guiding response systems and basic underlying limits in resources sufficient to maintain a prolonged response and match the scale of the biological threat itself expose capacity limitations [50,51].

Beyond planning and response resource capacities, population characteristics, and behavioral features of the greater Phoenix metropolitan area suggest additional challenges. A recent review of data gathered under the Centers for Disease Control and Prevention's National Immunization Survey shows that Arizona is in the bottom quintile of states in terms of vaccination rates in the United States [52]. This is a problematic indicator both in terms of overall resources in the region and from a behavioral perspective. Indeed, as earlier national efforts to inoculate health workers against smallpox risk was not particularly successful the context of low vaccination compliance or capacity raises another area of vulnerability [53,54].

Summary: Phoenix

Phoenix, and Maricopa County more broadly, is a metro area similar to the size of Sydney, approaching a population of five million people. A diverse and relatively economically prosperous area, it has significant resources at its disposal for the management of an epidemic. Potential adaptation to managing an epidemic is, in part, dependent on incident scale and duration—but also on the management of public resources in response to the outbreak. At the same time, Phoenix does seem to have nontrivial vulnerabilities rooted in a relatively poor state-wide track record of immunization and a relatively bifurcated age distribution along with somewhat limited hospital surge capacity. These and other characteristics, such as relatively low educational attainment for a significant proportion of the local population, would indicate vulnerabilities in the face of a smallpox epidemic. However, at the same time, overall community population density is relatively low for a population of its size and much like Sydney, relatively low-density housing stock might permit greater potential efficacy if household isolation is used as a risk mitigation strategy.

Analysis: comparison of the Sydney and Phoenix Areas

The following applies the determinants outlined in Table 10.3 to Sydney in the context of a hypothetical, large-scale outbreak of smallpox. An estimated maximum response capacity for the Sydney and Phoenix populations in response to an outbreak scenario occurring over a timeframe of weeks to a few months is outlined. For the purposes of comparison (Table 10.4), our analysis assumes that the initial phase of the outbreak would last no more than three months, beyond which time significant uncertainties exist regarding the longer-term response and adaptive capacity of these cities and the influence of the likely support that would be provided to these cities.

TABLE 10.4 Comparison of determinants of epidemic spread for Sydney and Phoenix.

Domain	Current analysis of Sydney and Phoenix	Estimated limit of adaptation for a major outbreak of smallpox
Interactivity	*Sydney:* High degree of interactivity; social distancing is achieved in many cases due to low population density; frequent mass gatherings; frequent minor gatherings; physically active population - recreational and leisure activities outdoor activities are a priority *Phoenix:* Social distancing not easily achieved due to population density; mass gatherings dependent in part on calendar timing; minor gatherings are frequent and not easily limited	Population is generally compliant with public health instruction at both locations – that potential for politicization is present and likely for Phoenix Social distancing is an inconvenience and does not impact on survival/ essential activities at both locations Loss of outdoor activities has significant cultural and mental health impacts at both locations
Flexibility and adaptability	*Sydney:* Cultural habituation to solitary activities and pursuits; generally low household size; infrequent intergenerational living arrangements; religious and cultural activities are rarely critically important *Phoenix:* Cultural habituation to individualized activities; low household size & infrequent intergenerational living arrangements; religious and cultural activities of moderate to significant importance	Widespread self-quarantine and self-isolation achievable and sustainable for both locations Essential workers are relatively easily segregated from rest of population for both locations Cancellation of cultural events generally acceptable with appropriate justification

Continued

TABLE 10.4 Comparison of determinants of epidemic spread for Sydney and Phoenix.—cont'd

Domain	Current analysis of Sydney and Phoenix	Estimated limit of adaptation for a major outbreak of smallpox
Transport	*Sydney:* Low to moderate reliance on public transport system; walking and cycling are common; individual motorized transport is dominant *Phoenix:* Minimal or very low reliance on public transport system; individual motorized transport is dominant	Most individuals have access to some form of individual transport in both locations. Traffic and congestion a concern for both sites — but more pronounced for Sydney if public transport not functional. Sydney's climate supports walking and cycling; extreme heat limits such options for personal transport in Phoenix for much of the year
Special groups	*Sydney:* Broad support services, delivered predominantly by strong state welfare system, to most vulnerable groups. *Phoenix:* Some support services made available to vulnerable groups with access and function needs, though generally not adequate to need	In Sydney, state welfare system may become overwhelmed in surge demand situation; likely also true in Phoenix In both locations, weakened system of community support may result in avoidable impacts on vulnerable and disadvantaged
Economic	*Sydney:* High income and significant disposable income; Many have significant savings; compulsory superannuation, with early access in emergency possible; high levels of household debt on average; low to moderate levels of government debt; market driven child care and aged care services, some relative disadvantages *Phoenix:* Uneven income distribution and limited slack resources for majority; significant	In Sydney, access to basic needs for survival secured for majority of population; greater disparities are present in Phoenix Child and aged care a problem should respective markets to fail

TABLE 10.4 Comparison of determinants of epidemic spread for Sydney and Phoenix.—cont'd

Domain	Current analysis of Sydney and Phoenix	Estimated limit of adaptation for a major outbreak of smallpox
	disposable income for portions of community; high levels of household debt on average	
Health — epidemic specific	*Sydney:* Federated-state based health systems; detailed pandemic preparedness and response plans; generic stockpiling of PPE and other essential items; public health specialist units; no specialized epidemic preparedness workforce on hand *Phoenix:* Government and employment-based health systems; detailed pandemic preparedness and response plans at national-level; limited, organization-specific stockpiling of PPE and other essential items; public health specialist units; limited specialized epidemic preparedness workforce	In Sydney, health department-led pandemic response involving whole-of-government coordination; Phoenix is more likely to utilized a unified command incident response structure with activation and release of strategic resources Repurposing of existing public health and hospital assets — rapid reorganization Deferral or cancellation of non-essential health activities
Health	*Sydney:* Moderate health resources by international standards per capita, particularly in key areas such as hospital beds, ICU beds & ventilators, and PPE stockpiles *Phoenix:* Moderate health resources by international standards per capita, particularly in key areas such as hospital beds, ICU beds & ventilators, and PPE stockpiles	Repurposing of existing resources to meet epidemic demand Rapid manufacturing and acquisition purchases Mobilization of workforce "Nationalization" of key personnel and assets Centralized control, decentralized execution

Continued

TABLE 10.4 Comparison of determinants of epidemic spread for Sydney and Phoenix.—cont'd

Domain	Current analysis of Sydney and Phoenix	Estimated limit of adaptation for a major outbreak of smallpox
Built environment	*Sydney:* Low density residences; some apartment living; high build quality, standards; low population density most of the time; culturally laissez-faire attitude to infectious disease prevention *Phoenix:* Generally low density residences with single family housing structure dominant; apartment living steadily increasing; high build quality, standards	In both locations, ability to separate and isolate high-density households into lower density units achievable Most households small in both locations Segregation in households more difficult to achieve Strong education required for inter-personal hygiene and contamination management
Workplace	*Sydney and Phoenix:* Most workplaces low density; some specific industries have high density crowded workplaces; face-to-face customer service roles are common; work from home or isolated not commonly available; culturally strong work ethos and presenteeism	Both locations have similar workplace profiles Work from home is possible Most individuals have technological options to enable remote work Social distancing achievable in many workplaces Open plan offices, service employment and similar are heavily impacted
Natural environment	*Sydney:* Mild to hot environment, with humidity; long distance travel normal *Phoenix:* Desert setting; hot to extended extreme heat affects personal mobility for significant parts of the year	For Sydney, environment not sufficiently hot or cold to prevent utilization of outside environment for travel, work or recreation For Phoenix, extreme heat does affect utilization of outside environment for travel, work or recreation—adaptations are required during parts of the year
Basic needs and essential services	*Sydney and Phoenix:* Well-resourced emergency response system across all sectors	In both locations, dependent on demand, emergency services may exceed surge capacity for periods of time

TABLE 10.4 Comparison of determinants of epidemic spread for Sydney and Phoenix.—cont'd

Domain	Current analysis of Sydney and Phoenix	Estimated limit of adaptation for a major outbreak of smallpox
	with some redundancy/surge capacity	Triage and rationing systems (already in place within policy) would be enacted Private system would have additional reserve capacity, but associated with cost
Waste and sanitation	*Sydney and Phoenix:* Sophisticated system in both sites; decontamination services are handled by fire and/or emergency medical or rescue service; potential for limited experience with biological hazard decontamination or management in given sites within broader metro areas	For both locations, absenteeism may impact on waste services and require specific management, but significant reserve capacity Additional decontamination resources could be surged from local defense forces and industrial organizations
Government	*Sydney and Phoenix:* Political system accustomed to emergency responses (bushfires, floods, etc.); agile emergency response systems in place already, but with some legacy structural inefficiencies; emergency powers and legislation well defined; review and audit of performance during emergencies are established, but insecure and open to political interference.	For Sydney, failure of state-based political leadership or operational functions mitigated with national- or other state-level capacity; similar intergovernmental challenges present in Phoenix Multiple simultaneous emergencies likely to overwhelm command and control system in both locations Accountability and transparency represent key concerns

Conclusion

An epidemic of any infectious disease with significant social and health impacts is a major challenge for a large city. Smallpox is an exemplar pathogen that has the capacity to cause catastrophic impacts on cities. Sydney and Phoenix have marked similarities in terms of population size, population density, residential features, population dispersal, transport distances, and key resources relevant to epidemic management. Geographic and unwarranted inequities and disparities across each city, political leadership and systemic inflexibilities, environmental conditions, and access to health care are key points of difference. Australia has universal healthcare insurance, whereas the United States does not. Lack of access to care as well as homelessness (present in both cities), race and ethnicity may result in worse pandemic outcomes [54]. More significantly both cities would need to significantly change and adapt both in the immediate response period, and likely in the medium- and longer-term. Planning for serial or multiple disasters would also improve resilience. It is also important to plan for intense outbreaks in vulnerable settings and populations — such as aged care, hospitals, homeless shelters, and prisons. The degree to which Sydney or Phoenix could adapt to a major outbreak of smallpox would be dependent on the baseline-level and effectiveness of preparedness and readiness activities, as well as the timeliness and effectiveness of system- and population-level adaptation and change. The recent outbreak of COVID-19 that has impacted both cities has demonstrated that rapid adaptation is possible to an emerging infectious disease crisis. However, it has also highlighted that response and adaptive capacity of both cities' systems may be insufficient particularly where competing interests are impacted as a consequence of epidemic response measures. Thus, certain factors similar to those uncovered during the COVID-19 pandemic could also jeopardize a response to a smallpox epidemic — especially poor or weak political leadership, resistance to population behavioral interventions, or the degradation of city resources and flexibility due to compound or complex events in addition to the outbreak.

References

[1] Sjoberg G. The origin and evolution of cities. Sci Am 1965;213:54—62.

[2] Bristow M, Bristow DN, Fang L, Hipel KW. Evolution of cities and urban resilience through complex adaptation and conflict resolution. Stockholm: Group Decision and Negotiation Conference; 2013.

[3] Thomas A. Urbanization before cities: lessons for social theory from the evolution of cities. J World Syst Res 2012;18:211—35.

[4] Dalziel B, Kissler S, Gog J, Viboud C, Bjørnstad O, Metcalf C, et al. Urbanization and humidity shape the intensity of influenza epidemics in U.S. cities. Science 2018;362:75.

[5] Sclar E, Garau P, Carolini G. The 21st century health challenge of slums and cities. Lancet 2005;365:901—3.

[6] Anderson R, May RM, editors. Infectious diseases of humans: dynamics and control. Oxford: Oxford University Press; 1991.

[7] Dembek Z, Pavlin J, Siwek M, Kortepeter M. Epidemiology of biowarfare and bioterrorism. In: Bozue J, Cote CK, Glass PJ, editors. Medical aspects of biological warfare. Fort Sam Houston: Borden Institute; 2018.

[8] Hassim M, Edwards D. Development of a methodology for assessing Inherent occupational health hazards. Process Saf Environ Protect 2006;84:378−90.

[9] Hassim M, Hurme M. Occupational chemical exposure and risk estimation in process development and design. Process Saf Environ Protect 2010;88:225−35.

[10] Koller G, Fischer U, Hungerbühler K. Assessing safety, health, and environmental impact early during process development. Ind Eng Chem Res 2000;39:960−72.

[11] Alhamdani YA, Hassim MH, Shaik SM, Jalil AA. Hybrid tool for occupational health risk assessment and fugitive emissions control in chemical processes based on the source, path and receptor concept. Process Saf Environ Protect 2018;118:348−60.

[12] McKone TE, Huey BM, Downing E, Duffy LM, editors. Strategies to protect the health of deployed U.S. forces: detecting, characterizing, and documenting exposures. Washington: National Academies Press; 2000.

[13] MacIntyre CR. Reevaluating the risk of smallpox reemergence. Mil Med 2020;185:e952−7.

[14] Noyce RS, Lederman S, Evans DH. Construction of an infectious horsepox virus vaccine from chemically synthesized DNA fragments. PLoS One 2018;13:e0188453.

[15] MacIntyre CR, Das A, Chen X, Silva CD, Doolan C. Evidence of long-distance aerial convection of Variola virus and implications for disease control. Viruses 2020;12:33.

[16] MacIntyre CR, Costantino V, Chen X, Segelov E, Chughtai AA, Kelleher A, et al. Influence of population immunosuppression and past vaccination on smallpox reemergence. Emerg Infect Dis 2018;24:646−53.

[17] MacIntyre CR, Costantino V, Mohanty B, Nand D, Kunasekaran M, Heslop D. Epidemic size, duration and vaccine stockpiling following a large-scale attack with smallpox. Glob Biosecur 2019;1:74−81.

[18] MacIntyre CR, Valentina C, Mohana Priya K. Health system capacity in Sydney, Australia in the event of a biological attack with smallpox. PLoS One 2019;14:e0217704.

[19] Kunasekaran MP, Chen X, Costantino V, Chughtai AA, MacIntyre CR. Evidence for residual immunity to smallpox after vaccination and implications for re-emergence. Mil Med 2019;184:e668−79.

[20] MacIntyre CR, Heslop D, Nand D, Schramm C, Butel M, Rawlinson W, et al. Exercise Mataika: White paper on response to a smallpox bioterrorism release in the Pacific. Glob Biosecur 2019;1:91−105.

[21] Australian Bureau of Statistics. 3218.0 - regional population growth, Australia, 2018−2019. Canberra: Australian Bureau of Statistics; 2019.

[22] Australian Bureau of Statistics. 3218.0 - regional population growth, Australia, 2018−2019. Canberra: Australian Bureau of Statistics; 2020.

[23] Australian Bureau of Statistics. Census of population and housing. Canberra: Australian Bureau of Statistics; 2016.

[24] Liu E, Easthope H. Multi-generation households in Australian cities. Sydney: Australian Housing and Urban Research Institute; 2012.

[25] CommSec. Australian home size hits 22-year low. Sydney: Commonwealth Bank of Australia; 2018.

[26] Australian Institute of Health and Welfare. Hospital resources 2017−18: Australian hospital statistics. Report No.: HSE 233. Canberra: Australian Institute of Health and Welfare; 2019.

[27] Barbieri S, Jorm L. Travel times to hospitals in Australia. Sci Data 2019;6:248.

[28] New South Wales Government. New South Wales state emergency management plan. Sydney: New South Wales Government; 2018.

[29] New South Wales Government. Biosecurity (animal and plant) emergency sub plan (version 5). Sydney: New South Wales Government; 2016.

[30] New South Wales Government. New South Wales human influenza pandemic plan. Sydney: New South Wales Government; 2018.

[31] New South Wales Health. HealthShare Sydney, Australia. 2020. www.healthshare.nsw.gov.au. [Accessed 26 August 2020].

[32] Arriagada NB, Palmer AJ, Bowman DM, Morgan GG, Jalaludin BB, Johnston FH. Unprecedented smoke-related health burden associated with the 2019-20 bushfires in eastern Australia. Med J Aust 2020;213:282–3.

[33] MacIntyre CR, Costantino V, Kunasekaran M. Health system capacity in Sydney, Australia in the event of a biological attack with smallpox. PLoS One 2019;14:e0217704.

[34] United States Census Bureau. Most of the counties with the largest population gains since 2010 are in Texas. United States Census Bureau; March 26, 2020.

[35] Arizona Department of Housing. Housing at a glance. Phoenix: Arizona Department of Housing; 2018.

[36] World Population Review. Maricopa county population. 2020. www.worldpopulationreview.com/us-counties/az/maricopa-county-population/. [Accessed 26 August 2020].

[37] Statistical Atlas. Maricopa county - age and sex. 2020. www.statisticalatlas.com/county/Arizona/Maricopa-County/Age-and-Sex. [Accessed 26 August 2020].

[38] Statistical Atlas. Maricopa county - race and ethnicity. 2020. www.statisticalatlas.com/county/Arizona/Maricopa-County/Race-and-Ethnicity. [Accessed 26 August 2020].

[39] Maricopa Association of Governments. Arizona demographics Arizona. Phoenix: Maricopa Association of Governments; 2020.

[40] Statistical Atlas. Maricopa county - household income. 2020. www.statisticalatlas.com/county/Arizona/Maricopa-County/Household-Income. [Accessed 26 August 2020].

[41] Arizona Department of Health Services. Arizona hospital facility ID list. Phoenix: Arizona Department of Health Services; 2020.

[42] Cohn AD. Mutual aid: intergovernmental agreements for emergency preparedness and response. Urban Lawyer 2005;37:1–51.

[43] Stier DD, Goodman RA. Mutual aid agreements: essential legal tools for public health preparedness and response. Am J Publ Health 2007;97:S62–68.

[44] Federal Emergency Management Agency. National incident management system. 2017. https://www.fema.gov/emergency-managers/nims. [Accessed 26 August 2020].

[45] Federal Emergency Management Agency. National response framework. 4th ed. 2020. www.fema.gov/emergency-managers/national-preparedness/frameworks/response. [Accessed 26 August 2020].

[46] U.S. Department of Veterans Affairs. VHA office of emergency management. National disaster medical system. 2020. https://www.va.gov/vhaemergencymanagement/. [Accessed 26 August 2020].

[47] Office of the President of the United States. National biodefense strategy. Washington: Office of the President of the United States; 2018.

[48] Arizona Department of Emergency and Military Affairs. Arizona state emergency response and recovery plan. Phoenix: Arizona Department of Emergency and Military Affairs; 2019.

[49] Government Accountability Office (GAO). COVID-19: opportunities to improve federal response and recovery efforts. Washington: Government Accountability Office; 2020.

[50] Propescu S. Arizona reopened too fast. Epidemiologists knew it, but we couldn't stop it. The Washington Post; July 16, 2020.

[51] Witte G, Weiner R. Sun Belt hospitals are feeling the strain from virus' surge — and bracing for worse. The Washington Post; July 16, 2020.

[52] Hess S. Arizona 8th Worst in U.S. when it comes to vaccination rates. Patch. January 21, 2020.

[53] Kuhles DJ, Ackman DM. The federal smallpox vaccination program: where do we go from here? Health Aff 2003;22:W3−503.

[54] Hsu HE, Ashe EM, Silverstein M, Hofman M, Lange SJ, Razzaghi H, et al. Race/ethnicity, underlying medical conditions, homelessness, and hospitalization status of adult patients with COVID-19 at an urban safety-net medical center - Boston, Massachusetts, 2020. MMWR 2020;69:864−9.

Chapter 11

The role of the private sector in urban health security

Irene Lai[1], Amy Simpson[1], Francesca Viliani[2], Philippe Guibert[3], Myles Druckman[4]

[1]Medical Information and Analysis, International SOS, Sydney, NSW, Australia; [2]Public Health, International SOS, Copenhagen, Denmark; [3]Europe Health Consulting, International SOS, Paris, France; [4]Global Health Services, International SOS, Los Angeles, CA, United States

The World Health Organization (WHO) has long recognized that global urbanization is a public health security threat [1]. As Thomas Bollyky states, "Health has shaped the history of cities, but it is cities that will define the future of global health and economic development" [2]. Crowded cities provide conduits for diseases to easily transmit to a larger number of people. Outbreaks can amplify and escalate rapidly in urban environments. The volume and speed of international trade and travel are contributing to the spread of communicable diseases. At the same time, cities could become a force for resilience against health threats through urban planning that includes health and social considerations as well as a network of social and economic forces that are active in its territory.

The outbreak of Severe Acute Respiratory Syndrome (SARS) in 2003 highlighted the impact a new pathogen could have on global health security. SARS became a disease of urban centers, where it most easily spread in city hospitals. It was easily spread between people, transcended geographical borders, and was quickly exported by air travelers around the world. Every international airport became a potential entry point. There was a significant toll on healthcare workers, travel and supply chains, as well as a large economic impact [1].

Prior to the Ebola outbreak in West Africa in 2014–16, the Ebola virus had mainly been found in rural settings in sub-Saharan Africa. But in late 2013, the virus made its way from a rural area in southeast Guinea into urban areas and across borders into neighboring Liberia and Sierra Leone. A small outbreak also occurred in Nigeria after an Ebola case arrived in Lagos, the most populous city on the African continent. Input from the private sector was identified as a crucial element in the response; private sector actors partnered

Inoculating Cities. https://doi.org/10.1016/B978-0-12-820204-3.00011-5
201

with the Lagos State Government to provide medical care and donate finances and equipment [3]. The WHO credits the Government's response, support from partners, and rapidly establishing an "Emergency Operations Center" for successfully averting a larger disaster [4].

This Ebola outbreak became the largest and most complex on record with over 28,600 cases in ten countries and took over two years to contain [5]. It demonstrated how lethal the virus could be in urban settings and the ease of international spread through highly mobile populations [6]. In addition, the epidemic was devastating for the socio-economic development of these countries [7]. Interestingly, it was not until 2015 that epidemics were characterized as disasters by the United Nations because of the negative impacts on development [8]. More recently, the spread of severe acute respiratory syndrome coronavirus 2 (SARS-CoV-2) and the resulting COVID-19 pandemic has demonstrated how wide and quickly a novel pathogen can spread [9]. Within a month of cases first being reported in the major urban center of Wuhan, China, the virus had spread to 18 other countries and the WHO declared the outbreak a Public Health Emergency of International Concern on January 30, 2020 [10]. The outbreak was characterized as a pandemic on March 11, 2020, by which time over 118,000 cases had been reported in 114 countries [11]. By the first of November 2020, over 46 million cases and nearly 1.2 million deaths had been reported globally [12].

Governments have a recognized responsibility for the health security of their citizens [13], in addition to providing access to safe water, sanitation, housing, and basic healthcare [2]. There are numerous opportunities for partnerships with the private sector and international organizations [14]. In the face of a crisis, the continuity of urban social structures and activities will be significantly affected by the response of private organizations. It is increasingly evident that the private sector has a significant role in health security beyond protecting the organization itself. The role of the private sector in health security is threefold − to safeguard its business and its stakeholders, protect and promote the well-being of its staff, community and clients, and support national and international public health efforts. This role is constantly evolving and at times extends beyond the direction of public health authorities. Three months after the outbreak of SARS-CoV-2 was characterized as a pandemic, the WHO issued a formalized request to the private sector globally and acknowledging that "the private sector has a critical role to play locally, nationally and globally" [15].

Infectious disease risks often pose a serious problem in the workplace, and as such, have always been a key concern of health and safety experts. Epidemics and pandemics have become an increasing concern for private companies.

Over the past 30 years, International SOS, a medical and security services company, has been involved in helping many leading domestic and multinational organizations from a wide range of sectors to prepare, plan and respond to health crises [16]. With operations and medical and security specialists in over 90 countries, International SOS has in-depth expertize in global crises and extensive experience in assisting governments, non-governmental organizations, and private companies in preparedness and response to outbreaks and pandemics [17]. This support has been not only in the provision of health advisory services to teams in charge of crisis management and business continuity but also to private sector leadership groups.

This chapter reviews national and international guidance and the evolution of private sector preparedness and involvement in health security. The urban environment is the focus as the private sector most readily functions out of urban environments, and thus has a vested interest in urban pandemic preparedness. Best practice principles for global health security planning and response for the private sector are provided. Finally, several case studies where International SOS has assisted organizations with addressing health risks, preparing pandemic and business continuity plans and providing other health advisory services to crisis management teams are presented.

A whole of society approach for pandemic preparedness and response

Health threats have proven time and again their ability to spread across borders and to be beyond the ability of any one sector to address them. In the 21st century, there is an evolving understanding that the private sector is critical in a whole of society approach to global health security. The International Organisation of Employers was one of the first networks of private sector organizations to liaize with international forums [18], such as the United Nations and the International Labor Organization.

Public-private partnerships for global health have been proliferating since the 1990s, when the 46th World Health Assembly called for "the support of all partners in health development, including non-governmental organizations and institutions in the private sector, in the implementation of their national strategies for health for all" [19]. The WHO further emphasized the need for "global cooperation in surveillance and outbreak alert and response between governments, United Nations agencies, private sector industries and organizations, professional associations, academia, media agencies and civil society" [1].

In 2009, the WHO put forward the Whole-of-Society Pandemic Readiness Framework for preparedness and response, encouraging private sector involvement, stating "Governments should actively promote the preparedness of the private sector" [20] particularly for pandemics. This approach emphasizes the interdependence of all sectors of society: government, civil society, and business. Nine key essential services are listed — health, defense, law & order, finance, transport, telecom, energy, food and water — all of which are supported by the private sector. In some locations, the private sector may be the sole provider of these essential services. The framework also recognizes that each sector has a role to play at different levels, including sub-national and local (i.e., urban). An all-of-society engagement and partnership to support disaster risk reduction is also one of the key pillars [8].

While industry associations and chambers of commerce can be part of national coordination bodies, the role that individual businesses play is crucial. Partnerships and alliances between public and private sector stakeholders are required to comprehensively respond to threats to health security, and the private sector has shown it is capable and willing to contribute to public health objectives. Over 9000 organizations have declared their commitment to the United Nations Global Compact "to align strategies and operations with universal principles on human rights, labor, environment and anti-corruption, and take actions to advance societal goals" [21]. Put forth in 1997 [22], and more recently reiterated by the United Nations Sustainable Development Goals (SDG), these partnerships should become the cornerstone of our collective response to pandemics and epidemics. Target 17.16 of the SDGs recognizes multi-stakeholder partnerships as important vehicles for mobilizing and sharing knowledge, expertize, technology, and financial resources [23].

The private sector must be included in planning and responding to health security threats in urban society. Society's health security cannot solely rely on government, even in jurisdictions that have robust public health infrastructure. Not only must the private sector support governments and communities, but it may also be called upon to lead. And while performing these roles, it has a duty of care to protect its own workforce and continue operations. It is however important to acknowledge that collaborations across sectors are not necessarily easy nor successful.

Epidemics are not different from other public health challenges as they are inevitably political issues [24]. The vested interest of the private sector should be acknowledged and addressed, as well as the importance of different skills and competencies. An example of private sector failure to assist with health security is the outsourcing of COVID-19 testing and tracing in the United Kingdom. Instead of being conducted by the National Health Service which has dedicated public health capabilities, the responsibility was given to private contractors [25]. At least three issues involving security, training, and faulty systems were identified and poor integration hampered response efforts [26]. At the time of writing, a parliamentary inquiry is underway in Australia into the government of the state of Victoria's response to COVID-19 [27]. After initially appearing to have controlled COVID-19 transmission in the state, a large second wave of infection originating in the state capital city of Melbourne led to the declaration of a state of emergency, hundreds of deaths, and extensive lockdowns across the state for almost three months. The outbreak has been linked to the hotel quarantine program for returning travelers. In particular, the utilization of private security services, rather than the Australian Defence Force or the State Police Department, has been highlighted as a contributory factor [28,29].

Organizational resilience

Collaborative approaches became an essential feature of business resilience to the risks posed by Ebola during the 2014—16 Ebola outbreak in West Africa [30]. Corporations realized that working independently was not offering protection to their investment and their people. Each epidemic is unique and comes with unique challenges. It is critical to ensure response systems evolve based on lessons learned from previous epidemics. Unfortunately, these lessons are primarily gathered within sectors with only limited learning on governance across sectors. Research about the role of the private sector in collaborating with other stakeholders in the management of disease outbreaks has shown that collaboration does occur between stakeholders, particularly in the mining sector [31].

This collaboration is ongoing during an outbreak, however, is rarely assessed, no collective learning is done and once the crisis is over each actor returns to its own comfort zone [32]. The main rationale for collaborations among multiple stakeholders was based on the high human, social and economic costs associated with uncontrolled epidemics. The private sector contributed knowledge, strategy and management value. Indeed, the Private Sector Roundtable emerged as a coalition of companies that acts as a central touchpoint for industry engagement to support countries in achieving the goals of the Global Health Security Agenda [33]. Following the success of the Private Sector Roundtable, the Connecting Business Initiative was established in 2016 [34]. It aims to integrate the private sector into national and international disaster management mechanisms as a means of reducing the risks and the duplication of efforts while increasing the resilience of businesses and society more broadly [35].

Private companies are often part of essential infrastructure and services for the functioning of urban society. Many are in the habit of identifying and managing risks as part of their business continuity strategy. It is also a key component of their duty of care toward their workers, and their social responsibility agenda coupled with expectations from stakeholders. Business continuity plans, for both private and public institutions, ensure that vital functions and services continue throughout an emergency. In fact, these plans are often required under local regulations by governing industry bodies and are considered a key dimension of planning and disaster preparation in cities and urban environments [36].

The workplace taking the lead in times of crisis

The COVID-19 pandemic has forced many companies to rethink their business and the relationship with their employees, their network of external collaborators, consumers, clients, as well as their surrounding communities. In the middle of this global crisis, we have seen a clear understanding that the response is a collective responsibility and the private sector needs to be part of the solution. Furthermore, the safe return of employees to work during the COVID-19 pandemic has been characterized as the moment when people and society try to regain control over the social and economic consequences of the pandemic. Safe work is, therefore, a key component of both business resilience and people's well-being.

Businesses can create a safe working environment, but as pandemics pose constantly evolving challenges, the solutions are based on interwoven responsibilities. Public health authorities provide measures and advice that need to be followed — employers have a duty of care to their employees and employees have a duty of loyalty to their employer (Fig. 11.1).

Briefly, duty of care refers to the moral and legal obligations of employers to their employees, contractors, volunteers, and related family members in maintaining their well-being, security and safety when working, posted on international assignments or working in remote areas of their home country. A company can implement risk reduction and preventive measures, but these must be adhered to by workers. Duty of loyalty from employees, refers to the adherence to all measures implemented by employers. Governments, at all levels, will continue to provide the regulatory framework for preparedness, response, and recovery and in some cases best practice framework for both operations and people. Organizations will need to adhere to the measures imposed by governments, while at the same time ensuring their business continuity.

In many cities, the workplace has become the center of attention — not only for direct employees but also for the community. Employees often turn to their employers for guidance and direction in the face of uncertainty and confusing direction from public health agencies and government. This has been further emphasized during the COVID-19 pandemic, where the response to the pandemic has been hampered by a corresponding "infodemic" or the rapid spread of information, inclusive of rumors, gossip and other forms of unreliable information [37].

FIG. 11.1   Interconnected duties of employers.

Organizations, therefore, play a crucial part in promoting and implementing health and infectious disease control activities — such as reinforcing the message that an employee who has been in contact with an infectious person self-quarantine for the recommended period. Rather than continue to reward the mentality that the most valuable employees ignore their own health needs, organizations need to promote a culture of self-care and protecting the health of fellow workers by not "soldiering on" when unwell.

Contact tracing — the process of identifying persons who may have come into contact with an infected person — is a vital element of urban health security. When applied systematically, contact tracing can break the chains of transmission of infectious diseases, and thus, is an essential tool for controlling infectious disease outbreaks [38]. In crowded urban environments, a single person may have dozens or even hundreds of contacts within a short period of time. The surge of COVID-19 cases among migrant workers housed in dormitories in Singapore is one dramatic example [39]. The quicker that contact tracing occurs and contacts are effectively managed, the better the chances are of limiting the size and duration of the outbreak. In many locations, the relaxation of movement restrictions applied to limit the spread of COVID-19 and enabling the resumption of community and social activities could only occur once contact tracing resources were scaled up. In some instances, this included digital apps that were developed through public-private partnerships [40,41].

Private sector businesses often need to support contact tracing efforts. For decades, airlines have been cooperating with health departments, providing information to assist with tracing contacts of travelers on identified flights. In locations where public health resources do not exist or are stretched beyond capacity, the private sector may help fill the gaps or complement efforts [42]. In addition, organizations may wish to conduct their own contact tracing of customers and employees as part of their duty of care. Several major corporations have developed internal protocols to carry out contact tracing within their workforce across their global premises in support and adherence to national efforts during the COVID-19 pandemic. These companies have therefore identified staff within the human resource and health and safety departments to carry out this new task and have trained them on how to do it while respecting the privacy of their workers. In urban office settings, organizations often keep track of employees and visitors to the premises through security swipe cards or a registration form. These logs can be used to determine the contacts of potentially infectious people. In the COVID-19 pandemic, these logs are encouraged or even mandated for some sectors in a number of countries. For example, dine-in restaurants can be required to keep a register of patrons with contact details and the date and time of dining. The list must be provided to public health authorities if requested.

The COVID-19 pandemic — a game changer

In a world with interlinked economic and social connections and totally interdependent supply chains, the role of business is central to infectious disease outbreak responses. This became clear in many different ways when SARS-Cov-2 began to spread around the world in early 2020, resulting in the COVID-19 pandemic. The WHO Global Preparedness Monitoring Board, in its first annual report published just six months earlier, warned "[the] world is not prepared for a fast-moving, virulent respiratory pathogen pandemic. [...] In addition to tragic levels of mortality, such a pandemic could cause panic, destabilize national security and seriously impact the global economy and trade" [43]. The response from many governments was slow and communications conflicting.

At the time of writing, these impacts are still playing out, and many of the examples cited are anecdotal. Some cities were unable to manage the thousands of people infected — hospitals, intensive care units, and morgues were stretched beyond capacity for weeks [44–46]. The impacts associated with public health measures — such as the implementation of cordon sanitaire, border closures, travel bans, prohibition of mass gatherings, shutdown of non-essential services and schools — include shortages of essential goods, the collapse of businesses [47], the stranding of travelers, and likely economic recessions [48].

For example, many governments, including some municipal governments, have mandated that travelers quarantine in designated facilities for 14 days upon arrival. International SOS has previously supported workplace quarantine programs in non-urban settings during the Ebola outbreak in West Africa in 2014–16 and the plague outbreak in Madagascar in 2017. It is our experience that during the COVID-19 pandemic, many private sector employers have designed their own quarantine programs for urban environments, in order to keep operating.

There are numerous examples of private sector players stepping in to overcome shortfalls during this pandemic [49–52]. In addition to the traditional involvement with donations of supplies and personnel — such as private health sector involvement [53,54], or pharmaceutical donations [55] — there have been many novel and innovative contributions. These range from supporting staff (e.g., the provision of accommodation and delivery of essential supplies to healthcare workers or vulnerable populations), making resources available (e.g., building temporary hospitals, innovative manufacturing of ventilators and testing kits, businesses diversifying or redirecting their capabilities to make hand sanitizer and producing personal protective equipment) and developing new technologies (e.g., up-scaling telemedicine, creating health code systems, contact tracing applications, providing platforms for online education during stay-at-home orders and screening tools for employees to return to work) [56].

Differing, conflicting, and confusing information has led to individuals and organizations looking beyond authorities for guidance. The discrepancy over school closures, with some facilities continuing face-to-face education, while others remain closed, has led to much confusion [57]. The differing and changing messaging about the use of face coverings and masks in public has resulted in organizations looking to trusted advisors to formulate their own policies and procedures. Plastic screens have proliferated in retail areas although there is no official guidance on the utility of these physical barriers. Many organizations have implemented entry screening, including a questionnaire and temperature measurement, although the practice has not been proven to reduce the risk of COVID-19 transmission.

Additionally, one downside of an open internet is that anyone can claim expertize. Indeed, the COVID-19 pandemic has been characterized by an overabundance of information that makes it difficult for people to find trustworthy sources and reliable guidance [58]. A Google search for "coronavirus" on March 27, 2020 generated nearly 10,500,000,000 results — many of these from people staying home in lockdown and reaching out to family and friends and the broader global community [59].

Without strong, clear leadership, social media and the internet can distort messages and cause confusion during a crisis; even in the best of times, conspiracy theories thrive. Organizations have had to take more control over health messaging and guidance for their workforce and community than ever before, to reduce the risk of inappropriate actions and reactions. In parallel with government-led responses, many organizations were assessing the same global information and actively formulating their response plans. The sentiment and actions taken by companies as the pandemic evolved were collected through polls performed by International SOS beginning on January 23, 2020. Over 800 companies and organizations participated. In this environment, polling has shown that employees will often look to their company for direction — particularly when their other authorities are perceived to be ineffective or biased.

In the week prior to the first poll, over 55% of participants were very or extremely concerned about the threat posed by COVID-19. The following week, 61% of companies were very or extremely concerned, and over half of the companies had banned travel to and from China. By mid-February, up to 67% were very or extremely concerned. By the time the WHO characterized the outbreak as a pandemic on March 11, 2020, many companies had already activated their pandemic plans. A poll conducted the week before the declaration revealed that 95% of companies had employees working from home, while of those companies with open operations, 65% were practicing social distancing.

COVID-19 also significantly impacted global organizations. The same mid-February poll revealed that nearly 70% of companies had employees quarantined while over 46% had employees that had become ill from COVID-19. By March 23, 98% of corporations were very or extremely concerned, and were planning for a prolonged and difficult challenge.

The structural changes around business operations in an urban setting during "lockdown" periods are likely to remain, at least in some format, until the pandemic ends. Working from home, where possible, has become the "new normal". However, there are many additional challenges and risks that organizations must address and support, including ergonomics, resilience, and mental health [60].

The importance of technology in a health crisis

The COVID-19 pandemic has occurred in a time when we have technologies available to efficiently engage and operate virtually. This has been a unique feature of this pandemic. Further, technology has not only changed the way we are doing business, but it may have saved many lives as well.

The WHO and Costa Rican Government have launched the COVID-19 Technology Access Pool which encourages solidarity and equitable sharing of data, scientific knowledge, technology, and intellectual property [61]. They are calling on multi-sectoral stakeholders to participate, including industry, researchers, funders, civil society, and governments. As the WHO Director-General Dr. Tedros Adhanom Ghebreyesus states, "Global solidarity and collaboration are essential to overcoming COVID-19" [62].

The pandemic has spawned many new technologies and pushed innovation to the forefront of many business operations. With government restrictions and limited movement, businesses have had to rapidly reinvent themselves, and find creative ways to deliver their services while keeping their employees and customers safe.

While video conferencing represents an obvious technology that has been critical during the COVID-19 pandemic, other less obvious tools have been created and are having an impact. Telehealth services have exploded. Family medicine cases are now being managed remotely via teleconference, and clinical decisions are being made without physical contact. Telemedicine will likely remain as a new and important healthcare option. Another example from healthcare is virtual reality technology that is being tested in COVID-19 intensive care units (ICUs) to limit the exposure of medical personnel to high-risk patients.

Technology has also allowed governments to share real-time information via email, website, or social media. Community-wide text alerts have become commonplace and important to inform the public of a critical situation. Academic and other expert organizations share and visualize analysis and data that previously would not have been available. These data can now be made available and shared with the private sector.

Organizations can now consolidate data from multiple sources and visualize this information against their work locations and other assets. They can then leverage medical and security expertize to rapidly communicate and respond to an evolving incident or crisis. All of this performed on a mobile digital platform that allows real-time reporting and monitoring of multiple threats on a global basis, at a granular, local level. What was once an almost impossible task, or would require dozens of staff, can be done with software tools. The raw data does require analysis and screening, and expertize is required to ensure its accuracy.

Apps developed by governments to support rapid national contact tracing are being adapted for organizational use. Once high-risk contacts are identified, companies use digital tools to monitor their quarantine staff, as well as screening them before entering the workplace (Figs. 11.2 and 11.3).

Pandemic planning beyond the healthcare sector

Organizations must ensure that their workplaces are safe from health threats. Occupational health and safety regulations and guidance related to infectious disease are generally aimed at occupations where exposure to pathogens is higher than during the course of daily living outside work. Outside of the healthcare sector, the risk to employees from infectious diseases present in the general community was not often included in traditional workplace hazard risk assessments.

Prior to the SARS crisis in 2003, traditional business continuity plans included scenarios for physical threats to assets — typically a bomb, fire, or

FIG. 11.2 Entrance screening of a major Australian bank, pandemic exercise 2009.

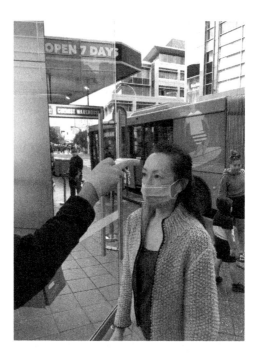

FIG. 11.3   Retail shop entrance temperature screening May 2020.

environmental disaster (e.g., a flood or cyclone), affecting only a single location or region. Few included any infectious disease, let alone pandemic scenarios affecting the entire organization in every location.

When the SARS crisis was declared over by the WHO in 2004 [63], this experience had highlighted to risk managers the global nature and pan-organizational threat of infectious diseases that spread from one human to another. Many regulatory bodies realized that the ability to respond to a pandemic — be it SARS, influenza, or another disease — was beyond the capabilities of governments and would require an all-of-society framework. Some have since mandated plans for businesses that perform or provide essential services.

An increasing number of authorities provide resources for businesses to plan for and respond to health threats. They are particularly relevant for operations in urban settings. These materials often had their genesis in advice for seasonal and pandemic flu, and have been updated and continue to evolve for the COVID-19 pandemic. Common areas included by authorities are screening and management of infected employees, cleaning and disinfection of premises, personal protective equipment, communication and education, and return to work.

At the time of writing, many of the general guidelines had been archived and replaced with advice specific to the COVID-19 pandemic. Examples of extensive guidelines specific to private sector businesses include: Enterprise Singapore's "Advisory on COVID-19 for businesses" [64], the Hong Kong Centre for Health Protection's "Guidelines" for Businesses & Workplaces [65], the International Chamber of Commerce's "Coronavirus Guidelines for Business" [66], the United Kingdom Government's "Working safely during coronavirus (COVID-19)" [67], the United States Centers for Disease Control and Prevention's Coronavirus Disease 2019 (COVID-19) "Businesses and Workplaces" [68] and the WHO's "Risk Communication for Employers and Workers" [69].

Many new triggering factors were also identified following SARS — particularly the role of international travel in the rapid spread of infectious diseases, the ability for pandemic threats to arise in any location, and the role of live animal markets, particularly in Asia, as the global epicenter of the human-to-animal interface. As a result, the urban environment quickly became a new key element in this equation. Firstly, the world had just passed the milestone of half of humanity living in cities in 2007, making this environment a priority and the next frontier to understand epidemiologic dynamics. Secondly, the world had been traumatized by the experiences of Toronto, Hong Kong, and Singapore — the cities most impacted by SARS and among the most densely populated places on earth — and their status as foyers of SARS transmission. Thirdly, it appeared that solutions to control the spread of diseases in urban settings had to be invented on the spot and had not been a part of any global or national pandemic preparedness plans.

Financial institutions were some of the early adopters of pandemic planning. Around 2007, organizations with major assets in urban settings in Asia assessed the potential impact of a pandemic, prompted by the H5N1 avian influenza virus. At the time, the virus was spreading rapidly around the world, causing extensive outbreaks in poultry in many countries and human cases were identified in over a dozen countries. The fear was, and still is, that H5N1 influenza might ultimately adapt to be highly contagious and transmissible between humans and hence ignite a severe pandemic. Given their proximity and geographical exposure, these organizations developed their initial influenza preparedness plans for a severe pandemic. The models assumed that the 20–40-year-old demographic would be the most affected, similar to the influenza pandemic of 1918, meaning that their workforce would be significantly impacted.

Initially, the topic of pandemic preparedness was new to many in most private sector organizations, which had virtually never been exposed to such events. The overall planning strategy was initiated by a few people, mostly security specialists; and the approaches used very much embraced a top-down model, with rigid plans to be implemented as they were created.

Lessons learned in the private sector from SARS

The first real opportunity to test the initial pandemic preparedness plans that were developed following the SARS outbreak arose during the 2009 H1N1 Influenza Pandemic [70].

What seems to have evolved over the years is based on the observation of several elements. First, there was a general acknowledgment that the detailed policies and procedures drawn up for a severe pandemic were incompatible with reality. Individual functions were required to operationalize and implement these plans. Many had aligned triggers with the WHO's influenza pandemic phases [71], which were not necessarily connected with the local epidemic situations. As a result, the actions were overly disruptive and many were not required for the relatively mild impacts of the 2009 pandemic. Plans were too rigid and unable to adapt to a less severe scenario with low absenteeism and low community anxiety, or the differing impacts from one location to another. Few had thought of identifying triggers for their company plans that would be independent of those used by international or national health authorities. The triggers to implement the elements of plans were based on factors that indicated an increasing risk of a pandemic at a global scale only, not at a country or even local scale.

A second aspect was related to those in charge of leading the private sector response. Risk management as a corporate function was in its infancy in many companies. Overall, the response to this event was the first opportunity for human resources practitioners to be invited to participate in pandemic preparedness and response. Individuals and employees were primarily recognized as the weakest link in the response to a crisis of infectious nature.

Subsequently, many of the private sector plans have been revised and broadened. They are now more flexible and have adopted an all-hazards approach to include threats other than infectious diseases, such as chemicals, radiation, air pollution, and other environmental risks. They also incorporate aspects such as continual risk assessment and surveillance, pharmaceutical and non-pharmaceutical measures, the use of digital tools, communications, education and awareness raising, and testing with scenarios and drills. The plans have demonstrated their usefulness as a basis for responding to a diverse range of scenarios, such as the Fukushima Daiichi Nuclear Accident in Japan 2011, the Western African Ebola virus epidemic from 2014 to 2016, and the rapid spread of Zika through the Americas from 2015 to 2016.

In 2009, many businesses were ready to shut down, which was not mandated by governments, and in retrospect, the plans were considered an "over-reaction." Paradoxically, for the COVID-19, based on the experiences a decade earlier, few anticipated the shut-down of transportation, borders, and businesses that have occurred.

In the past, private sector actors relied heavily on global and national health authorities to plan and react to health security threats. Now, and particularly in urban locations, it is imperative that all actors collaborate on their approach, planning, and response.

Best practice principles for global health security planning and response for the private sector

Based on International SOS experience and health authorities guidance [20,71−73], these principles will assist private sector companies that operate in cities and urban environments to prepare their pandemic and business continuity plans (Fig. 11.4).

There are nine components, with a crisis management team central to business continuity planning (Fig. 11.5). Seven outer elements then feed into the planning; surveillance, policies, liaison, risk assessment, infection control, active communication and resources (Fig. 11.6).

The following case studies were selected as they are representative of the connection between the private sector, preparedness, and response to health crises and urban environments. They describe how some major urban players, such as retail, utilities, and hospitality organizations, have prepared for and responded to pandemics. For each case study, references to the relevant elements of the best practice principles illustrated in Figs. 11.5 and 11.6 are provided.

FIG. 11.4 Best practice principles for health security planning and response for the private sector.

1. CRISIS MANAGEMENT TEAM
   a. Appoint a crisis management team that includes representation from every business line and takes into consideration the geographical spread of the company.
   b. Establish the chain of command structure, roles, and responsibilities.
   c. Appoint consultants to advise the team - trusted advisors and subject matter experts.
   d. Meet regularly, scale-up the frequency of meetings as threats arise.

2. BUSINESS CONTINUITY PLANNING
   a. Develop Business Continuity Plans which incorporate health considerations and pandemic specific preparedness.
   b. Identify the priorities of the company and the principles guiding and supporting the Plans, these will serve as an overarching framework for the more detailed actions.
   c. Develop specific plans for each business line and location, or a generic framework to be customized.
   d. Allow for a flexible scenario-based response, one that can adapt to a mild or a severe outbreak or health crisis, potentially affecting the organization at different magnitude in various locations, factoring:
      - Disease attack rate and case fatality rate;
      - Susceptibility of community or population;
      - Duration of the pandemic;
      - Multiple waves or phases; and
      - Government response and capability.
   e. Plan for recovery phase - ensure workplaces promote safe work practices that align with government advice, including social distancing and hygiene measures.
   f. Identify triggers for escalation and de-escalation.
   g. Identify critical roles - what supplies will they need, stockpiles of materials and equipment for work and health preservation.
   h. Consider how to support the local communities and actors in each location.
   i. Keep the plans alive through regular exercise, revise and adapt the plan, and report to executives.
   j. Ensure plans consider the return to operations and travel - taking into account local regulations, workplace risk, mitigation measures available, and the health of the workforce.

FIG. 11.5   Central elements of the best practice principles.

Duty of care case study: the hospitality industry

The hospitality sector is exposed to several challenges associated with ensuring their brand will still operate in the context of a crisis. This is particularly true during pandemic events, as the priority is to protect their employees, shift guests' behaviors toward safer alternatives, and adopt new protocols (e.g., enhanced cleaning measures).

The following example comes from a US headquartered hospitality group with global operations and their response to the 2009 H1N1 influenza pandemic. Their strategy to prepare for pandemic influenza was based on a policy issued by the company headquarters, which stated guideline principles and would need to be adapted and implemented locally by branches and franchises (Fig. 11.6, Item 4a). Ultimately, each regional country would

3. SURVEILLANCE
   a. Identify multiple reliable sources of information.
   b. Monitor the global, national, subnational and community threats and responses.

4. POLICIES
   a. Develop policies and procedures to minimize risk to business continuity and health impacts for employees, their families, and the community:
     • Communications – external and internal;
     • Information and Education – health messages;
     • Standards operating procedures (SOPs); and
     • Travel procedures.
   b. Consider:
     • The regulatory and ethical norms that the business operates in;
     • Inclusion of marginalized groups to avoid missing vulnerable individuals; and
     • Additional needs of essential workers; childcare, family support, psychosocial services.
   c. Cross train, especially for business-critical roles.

5. LIAISON
   a. Establish liaison points and collaborate with relevant groups at the local, national, and international levels. This is important for each company but is even more so for the essential sectors and critical infrastructure.
   b. Consider linkages and critical interdependencies between stakeholders (e.g., governments, regulatory bodies, business partners, suppliers, customers, employees, community groups)
   c. Share pandemic plans with partners and stakeholders to ensure harmonization, clarity, and consistency.

6. RISK ASSESSMENT
   a. Determine the events that can adversely affect the business and the possible consequences. Ensure continual surveillance of the global health threat landscape and ongoing dynamic risk assessments.
   b. Consider financial, security, infrastructure, and health risks. Include your supply chain and geographical variations

FIG. 11.6 Outer elements of the best practice principles.

develop its own plan, aligned with the guiding principles of the company. Three pillars did support planning: empowerment of managers, training of employees, and communication to guests. At headquarters, the focus was to get the latest updates on the global risk landscape, to adjust their response to provide support to affected areas (Fig. 11.6, Items 3a, 3b).

Some specific measures were considered to reduce the risk of transmission in hotels, resorts, and other premises. These included requiring waste management staff to use full protective equipment in the course of their duty and closing at-risk, communal places (e.g., recreation areas, swimming pools, and gyms).

7. INFECTION CONTROL

   a. Prepare an area where sick workers can be isolated until they can be taken to definitive care (or home if well enough) – a "temporary isolation room".

   b. Develop procedures and train personnel to manage illness in the workplace, including contact tracing.

   c. Consider non-pharmaceutical interventions:
- Social and physical distancing, working from home, telecommuting, etc.;
- Hygiene and cleaning;
- Isolation, quarantine, contact tracing;
- Screening;
- Workplace environment management; and
- Managing stress.

   d. Pharmaceutical interventions – access to prevention (vaccine if available) and treatments.

8. ACTIVE COMMUNICATION

   a. Have solid training and two-way communication workflows in place.

   b. Ensure ongoing communications with employees and other stakeholders throughout crises.

   c. Develop messaging and communicate transparently; include the known and the unknown. Incorporate uncertainty and the likelihood that information may change.

9. RESOURCES

   a. Have stockpiles of strategic supplies and equipment and allocate supplies to employees and customers as necessary.

   b. Ensure local internal leadership has the tools to communicate their status with senior leadership.

   c. Identify critical roles and functions, ensure back up resources are available and cross-trained.

FIG. 11.6 cont'd

Catering staff were further identified as a priority group. The organization took the opportunity to educate the catering staff on how to prevent transmission and increase overall hygiene standards and a series of posters were issued in several languages, covering activities in the following areas considered at-risk situations for company employees: reception, restaurant, conference rooms, shared areas, food preparation and handling, room cleaning, and laundry (Fig. 11.6, Item 8b).

Notably, as a follow-up to the preparedness plan issued for the H1N1 influenza pandemic, the group soon after considered it a priority to prepare for urban air pollution. This threat was indeed identified as material enough to represent a potentially significant disruptor to their business, though with a different pattern from a pandemic (Fig. 11.5, Item 2d). Cities where they operate, such as Beijing, Jakarta, and New Delhi, were regularly paralyzed by levels of poor air quality, potentially harmful to employee health and likely disruptive to business operations.

These episodes were considered to present a higher health risk to employees with specific medical conditions, such as heart or lung diseases. An Air Quality Contingency plan was modeled on the pandemic preparedness and response plan, with phases derived from outdoor air quality indices published by public health authorities (Fig. 11.5, Item 2f). Its objective was to ensure the employees' exposure to poor air quality was minimized, as a duty of care toward them.

The essential measures identified to gradually implement were self-identification of vulnerable employees, distribution of personal protective equipment, and working from home (Fig. 11.6, Item 9a). Other actions planned included central air filtration, maintaining a closed environment, limiting outdoor exposure and distribution of respirators. Finally, there was ongoing education and communication of travelers and visitors with regards to poor air quality events, including advising when to defer non-essential travel to affected locations (Fig. 11.6, Item 8b).

This hospitality group took advantage of both their pandemic influenza and air quality contingency plans as an overall matrix to respond quickly to the COVID-19 crisis. Anticipation of social distancing measures, an enhanced cleaning regimen, personal protective equipment stockpiling, and most notably, awareness of their staff to such health issues proved to be a competitive advantage for them to implement response measures swiftly.

Essential services case study: utilities

Urban settings are epicenters of utility services, given their concentration of inhabitants. The following example comes from a French group that provides utilities and waste management solutions globally. Their rationale for preparedness was based on the assumption that an influenza pandemic could infect 25%−30% of their workforce globally over a period of 12−15 months.

In 2006, motivated by fears that the avian influenza H5N1 virus may develop into the next pandemic, the group commenced a complete revision of their existing continuity plan. Indeed, this preparation allowed them to consider all aspects of business continuity, which could be impacted by different types of health disasters. This made it possible to deal with some complex problems in business like contagion, maintaining social distance, using protective measures, and addressing the issues of individual and collective hygiene at the workplace more generally.

Under the leadership of the Human Resources Department in Paris, preparedness was based on the innovative concept of describing all the granular aspects of pandemic response in a standard day of any employee (Fig. 11.6, Items 4a, 4b). Briefings, training, and communication would detail the behaviors expected by employees from the moment they would leave their home, commute, work, have meetings, break for lunch, and sometimes travel.

Each measure was supported by guiding documents, such as managing used masks, self-diagnosing symptoms of the flu, or traveling abroad in an affected country (Fig. 11.6, Items 7, 8c).

The risk management committee also created a pandemic flu simulation exercise kit intended to help the group's subsidiaries and their local teams to test their own level of preparedness (Fig. 11.5, Item 2i). A scenario of a mutated virus reported in a fictitious country would trigger the real-time management of the crisis when large uncontrolled clusters of transmission would be reported in a few cities where the group operated. Role definitions, scripts, and generic prompts were created with the intention of being customized by company offices in various countries. These exercises were not compulsory but were introduced as best practices to strengthen plans and teams. Several of them were conducted, mostly in Europe and in Asia, in the years preceding the H1N1 influenza pandemic.

Another focus of the plan was to maintain business travel for as long as possible in the event of a pandemic. The maintenance and servicing of utilities services in cities around the world were indeed a critical activity that could not be stopped during a crisis (Fig. 11.5, Items 2b, 2d, 2g). Dedicated features of the plan were created for business travelers, such as access to destination specific information on the impacts of the pandemic that would be available via an online travel portal. The Human Resources Management would have access to any traveler's location via a global travelers' locator platform and tailored travelers' kits containing a few facemasks, a thermometer, and hand-sanitizing gel would be available from all departure hubs (Fig. 11.6, Item 9a).

Overall, the response of this group to the H1N1 influenza pandemic in 2009 proved to be efficiently managed, with a mature, proportionate reaction to the evolving crisis. Business travel and domestic mobility were continued as much as possible, ensuring the presence of maintenance technicians in all relevant municipalities globally.

Essential services case study: the retail industry

Global and regional players in shopping centers and malls were quick to anticipate the impacts of a public health emergency on their businesses. Indeed, shopping center owners considered that given the large number of people visiting them on a daily basis, they would be immediately targeted by health authorities to reduce their activity, if not shut down completely, in an attempt to prevent mass movements of population and chains of disease transmission. Additionally, they feared shoppers would voluntarily avoid their premises once a new disease was declared transmissible between humans unless measures were taken to reassure them that the premises were safe and low-risk for transmission. In parallel, some centers were also recognized as potentially essential to their local community, as they had major food retailers and other tenants providing essential goods and services. A typically heavily regulated environment (e.g., hygiene, cleanliness, food supply, public space, etc.), they were confronted with new challenges inherent to preparing for a respiratory disease that spreads from human-to-human.

An example of this strategy comes from a large European player in the development and management of shopping malls. In 2007, this company based in the Iberian Peninsula had business continuity plans that did not include any health issues. The decision was taken by the board to assign the group Human Resources director as the leader for the revision of the plan and to prepare the company for an influenza pandemic (Fig. 11.5, Item 1a).

A pandemic preparedness exercise initiated the same year forced them to identify their weaknesses (Fig. 11.5, Item 2i). Within shopping centers, the company identified high-risk areas (e.g., restaurants, kindergartens, and fitness centers) and anticipated closing them, in line with guidance from the WHO [71] (Fig. 11.6, Items 6a, 6b). This exercise created an opportunity to define the most critical functions in shopping centers. Beyond the management team, roles such as cleaning services, security, waste collection, and maintenance were deemed essential.

From this planning exercise, it became obvious that roles such as cashiers, cleaners, or guards would become critical for the operation of shopping centers. Accordingly, these employees would require protection (e.g., shields, shifts, personal protective equipment, etc.) and training to keep the centers open (Fig. 11.6, Item 9c). As a result, the company considered stockpiling surgical masks and filtration masks for all their employees globally and have kept a minimum stockpile of masks ever since (Fig. 11.6, Item 9a).

Tenants were also considered a priority with regard to business continuity. Planning considered specific training for them to ensure they would adhere to a code of conduct allowing for safe shopping, and that they would install sanitation equipment in high-traffic, in-store areas. Training to educate shopping center managers and employees on the possibility of an emerging infectious disease became a part of the response strategy (Fig. 11.6, Item 8a).

Interestingly, no consideration was given to the maximum number of people allowed in a shopping center or store at the same time, which has become a key consideration during the COVID-19 pandemic.

Public-private partnership case study: the hotel industry

When New York City became a global epicenter of the COVID-19 pandemic in March and April 2020, it became apparent that the healthcare infrastructure was struggling to meet the rapidly expanding needs. Healthcare professionals (HCPs) were at an absolute premium to respond to the crisis. The Governor of New York urgently sought support from the private sector. A local hotelier offered its hotel in mid-town New York City to house HCPs who had volunteered to reinforce their medical colleagues on the frontlines who were falling ill or were burned out from work [74]. The hotel chain leadership, with the support and direction of International SOS, worked together to rapidly convert this luxury hotel into safe, tightly managed accommodation for medical personnel (Fig. 11.5, Item1c).

These arrangements required numerous town hall meetings to educate union employees and ensure they understood the risk and mitigation measures (Fig. 11.6, Items 5a, 5b). All hotel support personnel — including security, housekeeping, engineering, kitchen, and other hotel staff — were required to follow formal working procedures that emphasized social distancing and infection control precautions (including the proper use of personal protective equipment) as fundamental principles (Fig. 11.6, Items 8a, 8c). Once this was complete, the facility slowly opened its doors to the medical staff — increasing capacity by 10 people per day, with a maximum headcount of 150 persons — who were booked using a reservation system.

A single point of entry was created for both hotel employees and HCPs. This was staffed 24 h a day by nurses who conducted screening for everyone entering the premises, through a formal questionnaire and temperature checks. These nurses followed a bespoke screening protocol with an associated medical emergency response plan. The guest HCPs were given strict instructions on acceptable behaviors (e.g., no loitering, one guest per lift, one guest per room, etc.) and any violation would cause them to automatically forfeit their room (Fig. 11.6, Item 8b). Further, many of the communal areas in the hotel were blocked off. The HCP guests were not allowed to enter certain areas, to reduce potential exposure to hotel staff and maintain a clean guest space. The areas where the employees were segregated were outfitted with signage and hand sanitizers (Fig. 11.6, Item 7c). A "playbook" was created and distributed to hotel management that included policies, procedures and associated photos, posters and other communications tools.

The lessons learned and systems created during this project were subsequently implemented in a hotel in Boston the following week, where they were also instituted to support HCP.

This public-private partnership could not have been successful without the generosity of the private sector, the support of the local government, and on-site professional pandemic planning expertize (Fig. 11.6, Item 5b). This program ultimately protected not only the medical staff, but also minimized the potential spread to others in the community, including the family members of HCPs.

Conclusion

The importance of the urban private sector including health security as an essential element of their operational and business continuity plans is evident. Coordinated efforts across all industries are required to protect the health of the community. Increasingly, international and national authorities are adopting an all-of-society approach to planning for and responding to health threats. While the early engagement of the private sector was often limited to requests for funding and planning with the pharmaceutical industry, multi-country epidemics in the 21st century have highlighted the necessity of broadening the sector's involvement. The COVID-19 pandemic response and eventual post-pandemic reviews will no doubt serve as the basis of the framework for future threats.

References

[1] World Health Organization. A safer future: global public health security in the 21st century. Geneva: WHO; 2007.

[2] Bollyky TJ. The future of global health is urban health. Council on Foreign Relations; January 31, 2019.

[3] Otu A, Ameh S, Osifo-Dawodu E, Alade E, Ekuri S, Idris J. An account of the Ebola virus disease outbreak in Nigeria: implications and lessons learnt. BMC Publ Health 2018;18:3.

[4] World Health Organization. WHO declares end of Ebola outbreak in Nigeria. World Health Organization; October 20, 2014.

[5] United States Centers for Disease Control and Prevention. Ebola (Ebola virus disease): years of Ebola virus disease outbreaks. 2019. www.cdc.gov/vhf/ebola/history/chronology. html. [Accessed 13 June 2020].

[6] Eisenstein M. Disease: poverty and pathogens. Nature 2016;531:S61−63.

[7] Centers for Disease Control and Prevention. Ebola (Ebola virus disease): cost of the Ebola epidemic. 2019. www.cdc.gov/vhf/ebola/history/2014-2016-outbreak/cost-of-ebola.html. [Accessed 6 August 2020].

[8] United Nations Office for Disaster Risk Reduction. Sendai framework for disaster risk reduction 2015−2030. Geneva: UNDRR; 2015.

[9] World Health Organization. Rolling updates on coronavirus disease (COVID-19). 2020. www.who.int/emergencies/diseases/novel-coronavirus-2019/events-as-they-happen. [Accessed 12 June 2020].

[10] World Health Organization. Statement on the second meeting of the International Health Regulations (2005) Emergency Committee regarding the outbreak of novel coronavirus (2019-nCoV). World Health Organization; January 30, 2020.

[11] Ghebreyesus TA. WHO Director-General's opening remarks at the media briefing on COVID-19. World Health Organization; March 11, 2020.

[12] World Health Organization. WHO coronavirus disease (COVID-19) dashboard. 2020. http:// covid19.who.int/. [Accessed 5 November 2020].

[13] World Health Organization. International health regulations. 3rd ed. Geneva: WHO; 2005.

[14] United Nations Human Settlements Programme (UN Habitat). Annual progress report 2019. Nairobi: UN Habitat; 2019.

[15] World Health Organization. 'Asks' to the private sector in the response to COVID-19. World Health Organization; June 11, 2020.

[16] International SOS. Celebrating our 30th anniversary. 2015. www.internationalsos.com/ about-us/30th_landing. [Accessed 12 June 2020].

[17] International SOS. International SOS launches Enterprise health security center. International SOS; September 08, 2016.

[18] International Organisation of Employers. About us. 2020. www.ioe-emp.org/en/about-us/. [Accessed 12 June 2020].

[19] World Health Organization. Forty-sixth world health assembly – resolution WHA46, vol. 17. Geneva: WHO; 1993.

[20] Global Influenza Programme. Whole-of-society pandemic readiness: WHO guidelines for pandemic preparedness and response in the non-health sector. Geneva: WHO; 2009.

[21] United Nations Global Compact. Who we are. 2019. www.unglobalcompact.org/what-is-gc. [Accessed 12 June 2020].

[22] Kickbush I. New players for a new era: responding to the global public health challenges. J Publ Health 1997;19:171−8.

[23] United Nations. The sustainable development goals: sustainable development goal 17. 2015. https://sustainabledevelopment.un.org/sdg17. [Accessed 12 June 2020].

[24] MacPhail T. The viral network: a pathography of the H1N1 influenza pandemic. New York: Cornell University Press; 2014.

[25] Cave B, Kim J, Viliani F, Harris P. Applying an equity lens to urban policy measures for COVID-19 in four cities. Cities & Health; 2020.

[26] Pollock A. Thanks to outsourcing, England's test and trace system is in chaos. The Guardian; July 31, 2020.

[27] Parliament of Victoria. Inquiry into the Victorian Government's response to the COVID-19 pandemic. 2020. www.parliament.vic.gov.au/paec/inquiries/inquiry/1000. [Accessed 30 August 2020].

[28] Tobin G, McDonald A. Coronavirus quarantine guards in Melbourne hotels were recruited via WhatsApp, then 'told to bring their own masks'. ABC News; July 21, 2020.

[29] News.com Staff. Poorly trained security guards at hotel quarantine blamed for COVID-19 outbreak. August 04, 2020. News.com.au.

[30] World Economic Forum. Managing the risk and impact of future epidemics: options for public-private cooperation. Geneva: World Economic Forum; 2015.

[31] Dar OA, Viliani F, Tariq H, Buckley E, Omaar A, Otobo E, et al. Ebola and emerging infections: managing risks to the mining industry. In: Ali SH, Sturman K, Collins N, editors. Africa's mineral fortune: the science and politics of mining and sustainable development. London: Routledge; 2018.

[32] Viliani F, Edelstein M, Buckley E, Llamas A, Dar O. The infectious disease risk assessment and management (IDRAM) initiative: reflections from DRC pilot. Extr Ind Soc 2017;4:251−2.

[33] Global Health Security Agenda. Join the GHSA – Private Sector Roundtable. https://ghsagenda.org/joining-the-ghsa/. [Accessed 27 Oct 2020].

[34] Connecting Business Initiative. Engaging the private sector in disaster preparedness, response and recovery. 2019. www.connectingbusiness.org/about. [Accessed 12 June 2020].

[35] Connecting Business Initiative. Response to COVID-19: transforming private sector engagement in crises. Geneva: Connecting Business Initiative; 2020.

[36] United Nations Office for Disaster Risk Reduction (UNDRR). Disaster resilience scorecard for cities. Geneva: UNDRR; 2017.

[37] World Health Organization. Managing epidemics: key facts about major deadly diseases. Geneva: WHO; 2018.

[38] World Health Organization. Contact tracing in the context of COVID-19. Geneva: WHO; 2020.

[39] Leung H. Singapore was a coronavirus success story—until an outbreak showed how vulnerable workers can fall through the cracks. Time; April 28, 2020.

[40] Mozur P, Zhong R, Krolik A. In coronavirus fight, China gives citizens a color code, with red flags. New York Times; March 01, 2020.

[41] Apple Newsroom. Apple and Google partner on COVID-19 contact tracing technology. Apple; April 11, 2020.

[42] Bode M, Craven M, Leopoldseder M, Rutten P, Wilson M. Contact tracing for COVID-19: new considerations for its practical application. McKinsey & Company; May 08, 2020.

[43] Global Preparedness Monitoring Board. A world at risk: annual report on global preparedness for health emergencies. Geneva: WHO; 2019.

[44] Marks C. The wait is endless. Supplies are gone. My New York hospital is on the brink. The Washington Post; March 25, 2020.

[45] Sherwell P. Coronavirus: hospitals overwhelmed in the locked-down city of Wuhan. The Australian; January 26, 2020.

[46] Nacoti M, Ciocca A, Giupponi A, Brambillasca P, Lussana F, Pisano M, et al. At the epicenter of the Covid-19 pandemic and humanitarian crises in Italy: changing perspectives on preparation and mitigation. New Engl J Med Catal 2020. https://doi.org/10.1056/CAT.20.0080.

[47] Semuels A. As COVID-19 crashes the economy, workers and business owners wonder if anything can save them from financial ruin. Time; March 18, 2020.

[48] Organisation for Economic Co-operation and Development. Global economy faces a tightrope walk to recovery. 2020. www.oecd.org/newsroom/global-economy-faces-a-tightrope-walk-to-recovery.htm. [Accessed 12 June 2020].

[49] Knaus C. More than 130 Australian companies ready to boost PPE stock of coronavirus masks, gowns and gloves. The Guardian; March 23, 2020.

[50] Templeton B. Car companies are making ventilators, but ventilator companies, hackers and CPAP companies are working harder. Forbes; April 20, 2020.

[51] International SOS. COVID-19 client support at A glance, from international SOS. International SOS; April 15, 2020.

[52] Kelly J. The massive work-from-home COVID-19 test was a great success and will Be the new norm. Forbes; May 11, 2020.

[53] Hunt G. Australian Government partnership with private health sector secures 30,000 hospital beds and 105,000 nurses and staff, to help fight COVID-19 pandemic. Commonwealth of Australia, Department of Health; April 01, 2020.

[54] NHS England. NHS strikes major deal to expand hospital capacity to battle coronavirus. 2020. www.england.nhs.uk/2020/03/nhs-strikes-major-deal-to-expand-hospital-capacity-to-battle-coronavirus/. [Accessed 12 June 2020].

[55] Gilead. Gilead's investigational antiviral remdesivir receives U.S. Food and Drug Administration emergency use authorization for the treatment of COVID-19. Gilead; May 01, 2020.

[56] Tognini G. Coronavirus business tracker: how the private sector is fighting the covid-19 pandemic. Forbes; May 26, 2020.

[57] Blakkarly J. Teachers and parents slam mixed messages over school closures across Australia. SBS News; March 23, 2020.

[58] World Health Organization. 2019 novel coronavirus (2019-nCoV): strategic preparedness and response plan. Geneva: WHO; 2020.

[59] Viliani F. Infodemic ...We Need a Moment of Silence to Overcome this Pandemic!. LinkedIn; March 28, 2020. https://www.linkedin.com/pulse/infodemic-we-need-moment-silence-overcome-pandemic-francesca-viliani/. [Accessed 12 June 2020].

[60] Routley N. 6 charts that show what employers and employees really think about remote working. World Economic Forum; June 03, 2020.

[61] World Health Organization. International community rallies to support open research and science to fight COVID-19. World Health Organization; May 29, 2020.

[62] World Health Organization. Making the response to COVID-19 a public common good: solidarity call to action. Geneva: WHO; 2020.

[63] World Health Organization. China's latest SARS outbreak has been contained, but biosafety concerns remain – update 7. World Health Organization Disease Outbreak News; May 18, 2020.

[64] Enterprise Singapore. Advisory on COVID-19 for businesses. 2020. www.enterprisesg.gov. sg/esghome/covid-19. [Accessed 30 August 2020].

[65] The Centre for Health Protection, Department of Health. The government of the Hong Kong special administrative region. COVID-19 guidelines. 2020. www.chp.gov.hk/en/features/102742.html. [Accessed 30 August 2020].

[66] International Chamber of Commerce. ICC/NECSI publish business guidance to tackle COVID-19. International Chamber of Commerce; March 13, 2020.

[67] The United Kingdom Government. Working safely during coronavirus (COVID-19). 2020. www.gov.uk/guidance/working-safely-during-coronavirus-covid-19. [Accessed 30 August 2020].

[68] United States Centers for Disease Control and Prevention. Coronavirus disease 2019 (COVID-19): businesses and workplaces. 2020. www.cdc.gov/coronavirus/2019-ncov/community/organizations/businesses-employers.html. [Accessed 30 August 2020].

[69] World Health Organization. Risk communication. Empl Work 2020. www.who.int/teams/risk-communication/employers-and-workers. [Accessed 30 August 2020].

[70] Chan M. World now at the start of 2009 influenza pandemic. World Health Organization; June 11, 2009.

[71] World Health Organization. Pandemic influenza preparedness and response: a WHO guidance document. Geneva: WHO; 2009.

[72] World Health Organization. WHO pandemic phase descriptions and main actions by phase. Geneva: WHO; 2009.

[73] United States Centers for Disease Control and Prevention. Influenza (flu): global planning. 2016. www.cdc.gov/flu/pandemic-resources/planning-preparedness/global-planning.html. [Accessed 12 June 2020].

[74] Four Seasons. Four Seasons Hotel New York welcomes frontline healthcare heroes amid COVID-19 pandemic. Four Seasons Press Room; April 07, 2020.

Chapter 12

The health secure city: cities as conquerors of disease

Matthew Boyce, Rebecca Katz

Center for Global Health Science & Security, Georgetown University, Washington, DC, United States

In her book, *The Death and Life of Great American Cities*, Jane Jacobs writes that "Cities were once the most helpless and devastated victims of disease, but they became great disease conquerors" [1]. It was a combination of public health reforms, law, capacity building, infrastructure development, and scientific advances that led to improvements in urban health and allowed for cities to overcome more traditional scourges resulting from dense populations. For instance, in Stockholm, beginning in the mid-19th century, piped water, sewerage systems, and improved sanitation resulted in a remarkable decline in overall child mortality, and particularly, in mortality resulting from diarrheal diseases. As part of this effort, piped water was introduced to improve general hygiene, reduce the risk of epidemics, and enhance access to water; new sanitation ordinances were passed to increase the effectiveness and efficiency, the number of toilets greatly increased, and cesspools closed; the sewerage systems were altered to accommodate these changes, new plans developed, and wastewater treatment plans constructed [2]. Similarly, the provision of clean water in American cities – primarily through water filtration and chlorination – was responsible for reducing overall mortality by nearly half, child mortality by about two-thirds, and infant mortality by three-quarters [3].

These advancements likely reduced or eliminated the "urban health penalty" in many industrialized cities and are worthy of celebration. We now, though, find ourselves at a crossroads. Ahead of us lies a rapidly urbanizing world, where population growth in cities is overwhelming existing infrastructure and outpacing the development of new infrastructure – especially in cities in Africa and Asia. The provision of clean water and the construction of adequate housing and paved roads are not keeping up with the rate of urbanization in many developing cities, and local authorities are grappling with how to address these needs. Should they fail to do so, cities may once again become "devastated victims of disease."

Inoculating Cities. https://doi.org/10.1016/B978-0-12-820204-3.00012-7
227

This reality is coupled with new threats and challenges — such as climate change, globalization, and antimicrobial resistance. When combined with broader urbanization trends, these challenges could result in more infectious disease outbreaks in urban environments, affecting larger proportions of the population resulting migration, large scale outbreaks and pandemics, and an inability to effectively respond. A rather dystopic image.

Indeed, the COVID-19 pandemic has afforded a glimpse into what this future could look like whereby a novel pathogen emerges, rapidly spreads around the world as a result of global trade and travel, and local leaders are forced to mount a response hindered by high-level bureaucracy in the face of ambiguity and resource constraints. However, the pandemic has also afforded us glimpses at many potential solutions spearheaded by pragmatic cities, local governments, and municipal decision-makers.

A central theme throughout this book has been how cities and those with a vested interest in urban health have proactively prepared for infectious disease outbreaks. From the community-level action in Delhi to control mosquito populations and prevent dengue, to the vulnerability assessments in urban health systems in Phoenix, taking action to understand and mitigate the risks posed by infectious diseases before outbreaks occur has been common among the models discussed in the book. Much of this action was motivated by previous experiences with infectious disease outbreaks and there is little doubt that the COVID-19 pandemic will leave behind a legacy of preparedness.

Another theme has been the value added in exercising plans and conducting simulations. The mystery patient drills used by NYC Health + Hospitals allow for the city to exercise the existing plans, procedures, protocols and make adjustments as needed in healthcare facilities. Conversely, wargaming was shown to be a valuable simulation tool for proactively preparing — galvanizing discussion between various stakeholders who would be involved in the response to a public health emergency. After-action reviews represent another tool used in cities that hold immense value for assessing plans and how they were used in the response to actual public health events in urban areas as a means of improving future responses. We expect to see many of these following the pandemic.

Additionally, much as cities previously embraced technological advancements in the past to promote health, another recurring theme throughout the models discussed in this book is a willingness of cities to use new technologies. Kawasaki City's public-facing infectious disease surveillance system allows for rapid communication between health authorities in the city and affords residents of the city access to real-time surveillance data on the local level. While, across the world in the city of Blantyre, an innovative data pipeline is integrating varied data to inform and improve the local-level response to HIV.

And, whether it involves repurposing health infrastructure for broader pandemic preparedness, as is being done in cities across Myanmar or forming novel partnerships with private sector organizations, finding creative ways to address and overcome the crippling effects of resource limitations is a distinguishing characteristic of urban preparedness and response.

A new era of urban pandemic preparedness

If history repeats itself, as it often does, the time immediately following the COVID-19 pandemic will emphasize pandemic preparedness and the prioritization of public health. Some have already suggested that the post-COVID-19 world will be defined by a marked shift away from cities and the start of a new of de-urbanization trend defined by migration out of cities to suburbs and more rural areas [4]. Such a development would represent a major economic and cultural blow as cities represent some of the world's most efficient incubators for business and creativity. Others have argued that youth, eager to take advantage of opportunities missed during the COVID-19 crisis, will flock to cities once they are perceived as safe [5]. Regardless, following the pandemic, it will be paramount for cities and urban areas to prioritize pandemic preparedness and resilience. This would usher in a new era of urban pandemic preparedness.

The models discussed in this volume represent only a handful of tools in the toolbox of the innovative approaches being used in urban environments. Although these approaches are unlikely to work in all cities or urban areas — as the local contexts of cities around the world are richly diverse and "one size fits all" approaches that do not appropriately account for these differences are destined for failure — they represent powerful tools for ensuring that the health of urbanities, and our collective, global health is secure. Further, when used in combination with other tools and approaches — much as cities previously layered interventions to promote health and reduce the threats posed by infectious diseases — these models hold immense potential and could feature prominently in strategies for improving pandemic preparedness and response in urban environments.

City networks and innovative finance models will also be critical in this new era of urban pandemic preparedness. Cities have participated in networks for decades but recent years have witnessed an unprecedented proliferation of urban networks — largely the result of the regional integration processes of the 1990s and more nascent climate change movement that gained momentum in the 2000s [6]. The utility of these networks lies in their ability to provide a collective voice for influencing political agendas, as well as sharing resources — both material and intellectual.

Some estimates suggested that globally, over 200 formal city networks existed, with many more informal or ad hoc connections further defining the relationships between cities [7]. While the majority of these formal networks did not focus on public health, let alone pandemic preparedness, the COVID-19 pandemic caused many to pivot their focus, at least temporarily. This shift was catalyzed by the uninspiring reaction to the public health emergency by many national governments and an absence of practical guidance, coordination, and resources. The pandemic also spurred the creation of new networks focused exclusively on responding to the pandemic. This type of international collaboration and experience sharing among cities is different from more traditional multilateralism in that it reflects a type of global cooperation based primarily on pragmatism and problem-solving, rather than geopolitical interests [8]. Indeed, the pandemic has exposed many weaknesses in the multilateral system while concurrently highlighting the growing strength and dynamism of local leadership. While the sustainability of these pandemic networks is yet to be determined, and many of the existing networks that pivoted to pandemic response are likely to shift back to their initial or namesake causes, it is probable that at least a portion of these networks will continue to prioritize public health and preparedness. If this vision is realized, these networks will represent a new collective urban voice for improving urban pandemic preparedness. Longer term efforts in the broader multilateral context should embrace these organizations as partners and leverage them as a way of introducing new practical and direct perspectives, and work toward international goals, such as the Sustainable Development Goals or the objectives outlined in the World Health Organization's Global Vaccine Action Plan.

Of course, all of this will also come with a price tag, and as such, the financing of public health and pandemic preparedness in cities represents another crucial consideration. The enormous costs of infectious disease outbreaks are well documented, but governments with limited budgetary resources — a categorization that includes most, if not all, municipal governments — often fail to proactively fund measures to reduce the risk [9]. Instead, they tend to opt for more expensive response measures that are less effective from both health and economic perspectives. For example, the costs associated with containing the SARS-CoV-2 virus in Chicago exceeded 150,000 USD per week for the public health department alone in February of 2020 [10]. Furthermore, this cost estimate — already prohibitively high for many cities around the world, and incontestably unsustainable for many others — does not account for the costs associated with human illness and death, nor does it account for the indirect economic costs, such as associated with legally mandated business closures. Plainly, there is little question that preparing for infectious disease outbreaks is a more advantageous approach compared to responding to infectious disease outbreaks.

But cities and local policymakers are left with limited options for building the capacities required for financing these common goods for health and reducing the risks associated with infectious diseases. Traditional public financing models that blend revenues from local taxation schemes, support from central governments, municipal borrowing, and bond issues are often insufficient for developing the necessary capacities and sustaining them over long periods of time. And, in the wake of the COVID-19 pandemic, this will surely be true for cities who's bottom-line is determined by revenues generated from taxes [11]. Further, cities are generally able to access loans from multilateral development banks, such as the World Bank Group, only with the approval from their national government [12].[a] This frequently represents an insurmountable hurdle. As result, innovative financing models may also feature prominently in this new era of preparedness.

Two models that could prove especially practical for cities are social impact bonds and revolving funds. Social impact bonds, also known as pay-for-success contracts, are financial arrangements between public and private stakeholders that are earmarked for specific projects and tied to predefined outcomes [13,14]. Stated more simply, they are outcome-based contracts that leverage private investments to cover initial expenditures related to public goods. In a theoretical example, a city government, non-governmental organization or private sector actor, and investors would enter a contract in which the non-governmental organization or private sector actor would receive financing from the investors for a project to improve health with clearly defined objectives. If the objectives are met, the government would repay the investors their principal and a profit-margin determined by the level of success. In this theoretical, the city government would be willing to do so because improving health and pandemic preparedness saves the city money in the long run. However, should the stated objectives not be met, the investors would lose some or all of their money depending on the terms of the contract. These financing models were initially created to address problems that can be measured objectively, which could make it an especially good candidate for data-rich urban environments or for capacities that maintain the ability produce reliable data, such as infectious disease surveillance systems.

Revolving funds represent another innovative financing mechanism that cities could employee to improve urban pandemic preparedness. Under this model, a fund is created for a specified purpose with the understanding that repayments to the fund may be used again for the same purpose. Although recovering operating costs in early years can prove difficult and the repayment periods associated with these models typically require long periods of time, revolving funds offer several advantages in that once implemented, they are highly sustainable and the repayments can be used to finance additional projects as appropriate. Importantly, some precedent for this type of health financing already exists, as revolving funds have been previously used for water and infrastructure provision [15,16]. This mechanism could also provide a practical solution for financing pandemic preparedness in localities with underdeveloped credit markets but with the ability to enter into long-term contracts.

---

a. Some newer multilateral organizations have created mechanisms that allow for cities to bypass this obstacle and apply directly, but the effects of these have yet to be seen.

The health secure city

Cities around the world are treasures. They are centers of creativity, hubs of intellect, and economic powerhouses. But the COVID-19 pandemic has undeniably reminded us that, in our globalized world, cities can also pose unique challenges and act as incubators of disease — effectively pushing society to its limits. Still, the allure of cities remains and their future remains bright. And, much as cities have withstood the assaults of previous plagues, there is no reason to believe they will not also overcome this one. In fact, much as they have in the past, it is reasonable to expect them to responded by building back even better, safer, and more resilient than before.

In the long run, many of the models, tools, and approaches discussed in this volume will be essential for preparing cities for future epidemics and pandemics and for achieving a health secure urban environment. In the end, inoculating cities against future infectious disease threats.

References

[1] Jacobs J. The death and Life of Great American Cities. New York: Random House; 1961.

[2] Burström B, Macassa G, Oberg L, Bernhardt E, Smedman L. Equitable child health interventions: the impact of improved water and sanitation on inequalities in child mortality in Stockholm, 1878 to 1925. Am J Publ Health 2005;95:208—16.

[3] Cutler D, Miller G. The role of public health improvements in health advances: the twentieth-century United States. Demography 2005;42:1—22.

[4] Garrett G. The post-COVID-19 world will Be less global and less urban. Philadelphia: Perry World House; 2020.

[5] Florida R, Glaeser E, Sharif MM, Bedi K, Campanella TJ, et al. How Life in our cities will look after the coronavirus pandemic. Foreign Policy; May 01, 2020.

[6] Fernández de Losada A, Abdullah H. Rethinking the ecosystem of international city networks: challenges and opportunities. Barcelona: Barcelona Centre for International Affairs; 2019.

[7] Acuto M, Rayner S. City networks: breaking gridlocks or forging (new) lock-ins? Int Aff 2016;92:1147—66.

[8] Pipa AF, Bouchet M. How to make the most of city diplomacy in the COVID-19 era. Washington: The Brookings Institution; 2020.

[9] Peters DH, Hanssen O, Gutierrez J, Abrahams J, Nyenswah T. Financing common goods for health: core government functions in health emergency and disaster risk management. Health Syst Reform 2019;5:307—21.

[10] Harmon A, Stockman F. 'All hands on deck': health workers race to track thousands of Americans amid coronavirus. New York Times. February 22, 2020.

[11] Pagano MA, McFarland CK. When will your city feel the fiscal impact of COVID-19? Washington: The Brookings Institution; 2020.

[12] Hachigian N, Pipa AF. Can cities fix a post-pandemic world order? Foreign Policy; May 05, 2020.

[13] Warner ME. Private finance for public goods: social impact bonds. J Econ Pol Reform 2013;16:303—19.

[14] Edmiston D, Nicholls A. Social impact bonds: the role of private capital in outcome-based commissioning. J Soc Pol 2018;47:57−76.

[15] Holcombe R. Revolving fund finance: the case of wastewater treatment. Public Budg Finance 1992;12:50−65.

[16] Johnson CL. Managing financial resources to meet environmental infrastructure needs: the case of state revolving funds. Publ Prod Manag Rev 1995;18:263−75.

Index

Note: 'Page numbers followed by "f" indicate figures and "t" indicate tables.'

Printed in the United States
by Baker & Taylor Publisher Services